Groundwater - Resource Characterisation and Management Aspects

Edited by Modreck Gomo

Published in London, United Kingdom

IntechOpen

Supporting open minds since 2005

Groundwater - Resource Characterisation and Management Aspects
http://dx.doi.org/10.5772/intechopen.73345
Edited by Modreck Gomo

Contributors
Jose Luis Arumi, Enrique Muñoz, Ricardo Oyarzun, Olena Vavrinevych, Anna Antonenko, Sergiy Omelchuk, Maria Korshun, Dalia Elsheakh, Esmat Abdallah, Abhay Soni, Oscar Alfranca, Miha Curk, Matjaž Glavan, Joe Magner, Modreck Gomo

First published in London, United Kingdom, 2019 by IntechOpen
IntechOpen is the global imprint of INTECHOPEN LIMITED, registered in England and Wales, registration number: 11086078, The Shard, 25th floor, 32 London Bridge Street
London, SE19SG - United Kingdom
Printed in Croatia

British Library Cataloguing-in-Publication Data
A catalogue record for this book is available from the British Library

Additional hard copies can be obtained from orders@intechopen.com

Groundwater - Resource Characterisation and Management Aspects
Edited by Modreck Gomo
p. cm.
Print ISBN 978-1-83880-514-2
Online ISBN 978-1-83880-515-9
eBook (PDF) ISBN 978-1-83880-711-5

For EU product safety concerns:
IN TECH d.o.o., Prolaz Marije Krucifikse Kozulić 3, 51000 Rijeka, Croatia, info@intechopen.com or visit our website at intechopen.com.

Meet the editor

Modreck Gomo is a researcher in groundwater with experience in a wide range of groundwater science and consultancy projects. His first degree was a BSc Honours in Agricultural Engineering from the University of Zimbabwe, followed by a BSc Honours in Geohydrology (2008), an MSc in Geohydrology (2009), and a PhD in Geohydrology (2011) from the University of the Free in South Africa. He has published over 20 research papers in peer-reviewed international journals.
His main research interests include:

- Investigation of the characteristics of borehole fluid electrical conductivity (FEC) profiles in different aquifer systems;

- Application of numerical modelling to assess and evaluate the application of analytical methods for analyzing aquifer pumping test data;

- Investigation of hydrogeochemical processes and groundwater quality assessments; and

- Procedures for collecting representative groundwater samples in fractured-rock aquifers.

Dr Gomo is a consistent reviewer in more than 10 internationally peer-reviewed journals with a focus on groundwater science since 2012. He is also now involved in a number of groundwater projects in the Southern African Development Community region related to transboundary groundwater and capacity building on groundwater collection. He is a registered professional natural scientist in the field of water resources with the South African Council for Natural Scientific Professions (SACNASP) and also a member of the Groundwater Division, South Africa.

Contents

Preface

In practice, most of the world's aquifers are inherently heterogeneous and isotropic in nature. This calls for continuous improvement and understanding of hydrogeology site characterisation techniques in theory and application to better understand and manage groundwater. The effects of climate change have in some cases resulted in unpredictably prolonged and frequent droughts, which have brought stress to groundwater resources. The effect often translates into unpredictable borehole yields and sometimes the drying of boreholes.

This book is structured into two sections that cover aspects of hydrogeology site characterisation and groundwater management.

Hydrogeology Site Characterisation

The first part of this section presents the principles and applications of ground penetrating radar (GPR) as a promising technology to detect and delineate aquifers or non-metallic minerals. Due to the inherent heterogeneity and isotropic nature of the subsurface, the need to improve approaches for groundwater detection and delineation remains vital.

In the second part, a case study on the hydrogeology characterisation of the Andean mountain groundwater resources of Central Chile is presented. Field surveys, streamflow gauging, environmental tracers, and a hydrological model are utilized as complementary techniques to improve site understanding.

Methods for predicting the risk of groundwater contamination with pesticides and their dangerous aspects for human health are then developed. These methods comprise risk acceptance assessment and integral groundwater contamination hazard index (IGCHI) evaluation according to a special scale.

Groundwater Management

The first part of this section discusses the sustainability of human, plant, and aquatic life influenced by groundwater systems from recharge to discharge. Considerations are given to current and past stressors of groundwater by using case examples from around the world. In doing this, hydrogeologic settings where anthropogenic activity has impaired or has the potential to impair human, plant, and aquatic life are explored. The second part of the section addresses the management issues of mine groundwater quantity and quality during and after mining. A good understanding of the interrelationships between groundwater hydrology and mining processes is important to sustainably manage mine water. This enables efficient planning of mining processes and water management to protect the groundwater while addressing water scarcity and security issues in communities within the vicinity of mining areas.

The third part of the section presents legislative aspects to protect groundwater as instruments of groundwater management using a Slovenian example. To achieve this, an overview of groundwater protection practices in Slovenia is first presented. The case studies evaluate the "theory" by reviewing national legislations of concern. The practice looks at the guidelines and solutions drawn from legislations to comply with the European Union Water Framework Directive (WFD). Furthermore, a discussion of the current activities aimed at improving Slovenia's groundwater status is given. The validity of the Gisser-Sanchez model (GSE) hypothesis in groundwater management is re-evaluated and examines the conceptual framework within which the elements interacting in the management of groundwater resources are examined. The section concludes that the role of the market is limited with respect to the price of water in an aquifer.

We believe that this book can be useful for various professionals involved in groundwater-related work to improve the theoretical and practical understanding of hydrogeology site characterisation techniques and groundwater resource management skills.

Modreck Gomo
Institute for Groundwater Studies (IGS),
University of the Free State,
South Africa

Section 1

Hydrogeology Site Characterisation

Chapter 1

Detection of Underground Water by Using GPR

Dalia N. Elsheakh and Esmat A. Abdallah

Abstract

Water is the human vital requirement for life; in these days, decreasing of the fresh water increases the importance of the aquifer water. However, Upper Egypt is higher than north Egypt, so the water map continually changes daily, and the aquifer water is deeper than 10 m. The ground penetrating radar (GPR) system is used for underground water detection. GPR is a promising technology to detect and identify aquifer water or nonmetallic mines. One of the most serious components for the performance of GPR is the antenna system. The technology of the remote sensing and radar is rapidly developing, and it has led to the ultra-wideband electronic systems. All of these factors, such as miniaturized, low cost, possible compromise solution between depth and resolution, scanning in real time, easy to interpret, and decreased the false alarm, are important in designing the ground penetrating system. The electrical properties of the sand and fresh water layers are investigated using laboratory measurement and EM simulation. Different types of antenna may be used in GPR to operate over a frequency range for different penetration depth. Frequency-modulated continuous wave is also used for GPR and for through-the-wall applications. However, most of these kinds of antennas are limited by their large volume for certain applications. Therefore, a compact Vivaldi antenna with EBG and a compact planar printed quasi-Yagi antenna with meandered ground plane are designed to fulfill all above requirement.

Keywords: high-frequency structure simulator (HFSS), Yagi antenna, Vivaldi antenna, ultra-wideband (UWB), ground-penetrating radar (GPR), water detection, printed antenna, reflection coefficient, phase

1. Introduction

After the year 2002, ultra-wideband (UWB) systems have gained popularity mainly when the US Department of Federal Communications Commission (FCC) allocated a license-free spectrum for industrial and scientific purposes. FCC is doing the greatest step in opening new doors of researches for UWB in the field of wireless communications and microwave imaging [1, 2]. UWB device is defined as any device operating in absolute bandwidth greater than 500 MHz or fractional bandwidth greater than 0.2 of central frequency [3]. The frequency band ranges of UWB extended from 3.1 to 10.6 GHz that have expected the applications in the fields of wireless body area networks (WBAN), wireless local area networks (WLAN), wireless interoperability for microwave access (WiMAX), wireless personal area networks (WPAN), and ground-penetrating radar (GPR) technology where wide bandwidth is required [4]. GPR is the major applications of UWB technology, which is in large degree used in military and civilian applications such as water detection and land mines [5]. There are many UWB antennas have been designed for GPR applications. The study based on the lower-frequency band is conducted mainly to increase the penetration depth, while the designing in the higher-frequency band is performed to achieve high-resolution imaging for GPR systems. Some of the researches focused on the entire UWB frequency range to further improve the bandwidth, while others focused on enhancement of the antenna gain [6]. Moreover, GPR is also used in remote-sensing techniques as nondestructive testing of concrete and detection of trapped people under debris or in opaque environment [7]. For the achievement of UWB GPR systems, the performance of various antenna designs, such as bow-tie antenna [8], spiral antenna [9], loaded dipole antenna [10], TEM horn antenna [11], tapered slot antenna (TSA) [12, 13], and Vivaldi antenna [14, 15], has been evaluated.

Ultra-wideband antenna is one of the preferred antennas for some applications in the microwave imaging, object measurement technology, and noninvasive testing (NIT) [4–15]. In this era, the microstrip antenna has the advantages of small size, high gain, and low cost for good performance in some applications [6–10]. Ground-penetrating radar (GPR) has been utilized by emitting an electromagnetic wave directed into the ground, and the buried objects cause reflections of the emitted wave that are then detected by the receiver system. This is contrasted in the electrical properties as the signal reflection coefficient and their related phase [11–17]. GPR has an ability to detect the electrical inhomogeneity of metal and dielectric object in the presence of surrounding soil or sand [18]. GPR system can exist at the same location for the transmitting and receiving, and there are four types of GPR system: quasi-monostatic radar if there is no separation distance between transmitter and receiver, monostatic radar if there is single antenna performs both transmit and receive operations [19], bistatic radar if the transmitter and receiver have separate distance, and multistatic radar if a radar system involves one or more transmitting platforms and multiple receiving platforms [20]. GPR systems have been classified as the time domain (impulse radars) and continuous wave (CW) radar [21]. CW radar transmits the signal, which can be frequency-modulated continuous wave (FMCW), or creates the resulting signal as a combination of monochromatic steps through a certain band of frequencies, referred as stepped frequency continuous wave (SFCW) [22]. GPR systems usually work at central frequencies below 1 GHz, and large bandwidth is needed for a better depth resolution and detailed echo. The use of impulse wideband systems involves some technical problems, such as Doppler processing, propagation fading, interference rejection, wave clutter, detecting birds on or near the water surface, and radar interference.

2. GPR system

Egypt desert water, land mines, and Egyptian ancient mummy detection are an important, and yet challenging problem remains to be solved and is a matter of concern for both civilian groups and military. Water depth, quantity, and volume are different from location to another, and landmines have huge variety in use with various sizes and materials. They are mostly designed in a circular, rectangular, butterfly, or cylindrical shapes as V69, PSM-1, Hamdy, and Mon 200 with dimensions 13, 7.6, 21, and 43 cm, respectively. On the one hand, the electrical properties of water purity are changed due to water quantities and mineral contained; on the other hand, the electrical properties of the mummy are varying according to biological materials involved as tissues, bones, scapula, and femur, respectively [23]. The average effective dielectric constant is equal to 6.5 and dielectric loss equal to 15 [24].

Groundwater is a hidden natural resource. It is found in different proportions, in various rock types, and at various depths of ground surface. Previously, in the past, when there is no visible flow of water along the rivers or lakes, people used to dig small pits, in the river alluvium, then collect the groundwater coming through seepage to use in different purposes and for meeting the domestic needs. As well as the people of mountainous area, springs are the outcome of seepage from any groundwater system [25, 26].

More than 60% of the global population thrives by using only the groundwater resources. The groundwater, which was existing at shallow depths in the open wells, has gone deep due to overexploitation. Exploring these water sources becomes a challenging task to geoscientists [27–29].

As the signal of the GPR is required to propagate through inhomogeneous media, the efficiency of the antenna should be taken into account. Thus, in order to enhance the efficiency, a compact shape of Vivaldi is designed [7]. The compactness in the shape of antenna is achieved by using two Vivaldi surface shapes with the same feeding network. Frequency-modulated continuous wave (FMCW) is another type than pulse wave also used for GPR and for through-the-wall applications. FMCW systems are transmitting a repetitive waveform with increasing or decreasing frequency [21]. UWB-FMCW systems require a linear sweep, which may be difficult to create. Lower-frequency signals are required to penetrate the ground effectively as the penetration depth decreases with increasing frequencies.

GPR radar could be used for soil with different electrical properties to detect water underground [30, 31]. However, a broadband signal with good resolving power is required. Thus, the GPR antenna should operate on lower frequency and should have a very wide bandwidth. Also, the UWB antenna applied to communication system with high gain is highly desired [32].

3. Compact ultra-wideband (UWB) Vivaldi antenna

The main object of this section is the detection of water. A compact novel shape of Vivaldi antenna as shown in **Figure 1** with dimensions of $0.17\lambda \times 0.16\lambda \times 0.013\lambda$ is proposed in this section [6].

A comparative study was undertaken as shown in **Table 1** for different Vivaldi antenna that are used of GPR application in the same interesting operating frequencies [8–11] to show that the proposed antenna gives compact size with higher gain from 250 MHz up to 10 GHz.

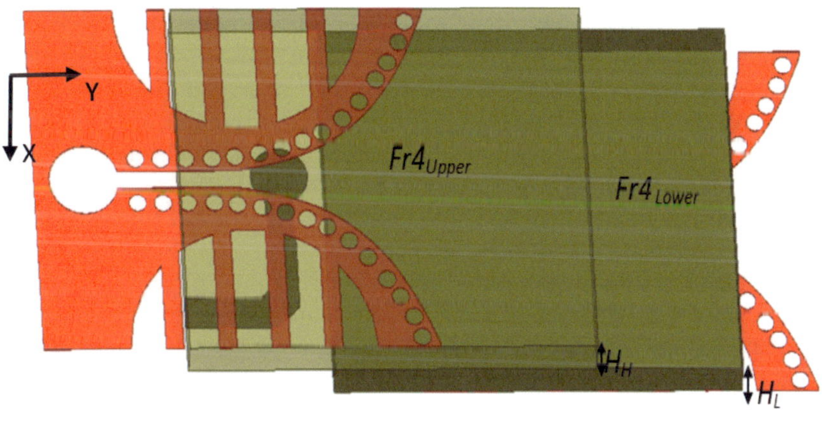

Figure 1.
3D geometry of the proposed Vivaldi antenna.

Ref.	L × W cm^2	Sub. thickness (cm)	Diel. properties	Gain (dBi)	BW (MHz)
[8]	60 × 40	0.32	εr = 4.4, tanδ = 0.02	9	500–1500
[9]	60 × 30	0.315	εr = 2.33, tan δ = 0.002	9	500–1500
[10]	7.8 × 7.5	0.16	εr = 4.4, tan δ = 0.02	8	1000–4000
[11]	100 × 90	1	εr = 4.4, tan δ = 0. 02	10	50–250
Ours	13 × 12	2	εr = 4.4, tan δ = 0.02	17	250–10,000

Table 1.
Comparison of the proposed antenna with other antennas (all dimensions in cm).

There are a number of techniques used for improving the parameters of printed antennas with different feeding techniques. These techniques include electromagnetic bandgap (EBG), metamaterial, and defected ground structure (DGS). EBG structure has gained popularity among all the techniques reported for enhancing the parameters due to its simple structural design. The periodical shapes are etched as square, mushroom, and circular shapes in the radiator or ground plane to achieve inductive and capacitive load to create band-stop characteristics and to suppress higher mode harmonics and mutual coupling.

The basic concepts, working principles, and equivalent models of the different shapes of electromagnetic bandgap structure (EBG) are presented [36]. EBG has been used in the design of the Vivaldi antennas for improving the bandwidth and gain of proposed antenna and suppressing the higher harmonics mode and mutual coupling between adjacent elements. In addition, the proposed antenna cross-polarization is improved for the radiation characteristics [6].

3.1 Vivaldi antenna geometry and principle theory

A relatively large number of published UWB Vivaldi antennas consist of a feed line and exponential ground plane. Our proposed antenna starts as shown in **Figure 2(a)** from one layer of FR4 low-cost substrate with conventional exponential Vivaldi tapered slot line shape using empirical Eq. (1) [4]:

$$y = \pm 0.018\, e^{0.27x} \tag{1}$$

where x and y are the axes of the inner and outer exponential to improve the impedance bandwidth. The first step is without any slot, **Figure 2(a)**, the second with circular EBG-etched slot on the edge of exponential tapered slot.

The symmetrical circular electromagnetic bandgap structure (EBG) slots are etched to increase the antenna bandwidth as shown in **Figure 2(b)**. The modified version from the first one is obtained by slotting the two arms of the Vivaldi antenna.

To improve the bandwidth of the Vivaldi antenna, symmetrical semicircular slots are etched to increase the antenna bandwidth as shown in **Figure 2(c)**. **Figure 2(c)** and **(d)** shows the geometry of the proposed antenna in two different configurations. The substrate used is FR4 dielectric substrate of thickness 1 cm, a relative permittivity of 4.4 and a loss tangent of 0.02.

Finally, dual FR4 substrates printed with Vivaldi ground plane with the same feeding line are used. The final geometry of the proposed antenna design with ground plane and feeding network is shown in **Figure 3**. This design is used to investigate a dual substrate layer of Vivaldi ground plane antenna as shown in **Figure 3(a)** with the same feeding network as shown in **Figure 3(b)**. Proposed Vivaldi antenna consists of two layers from the dielectric sheet, the feeding line is sandwiched between them, and the two layers of metallic Vivaldi are mounted on

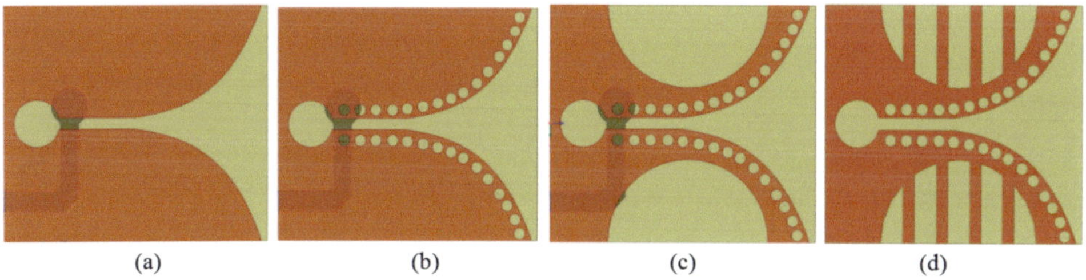

Figure 2.
The top view of Vivaldi antenna, (a) conventional Vivaldi, (b–d) modified Vivaldi antenna.

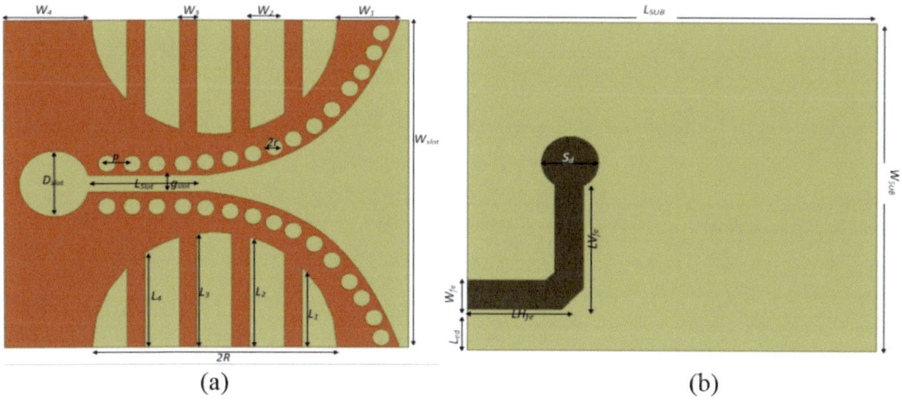

Figure 3.
Geometry of the proposed Vivaldi antenna (a) ground plane and (b) feed line.

the dielectric substrate (one at the top and the other at the bottom) as shown in three dimensions of the proposed Vivaldi antenna in **Figures 1** and 7.

3.2 Antenna simulation and measured results

The reflection coefficients of the antenna versus frequency for the four-step design of Vivaldi antennas in frequency range from 0.25 to 10 GHz are shown in **Figure 4(a)**, and the zooming range from 0.25 to 2 GHz is shown in **Figure 4(b)**. The final antenna design achieves improvement in antenna impedance matching all over the band. The dimensions of the proposed antennas are shown in **Table 2**.

The surface current density distributions of the compact Vivaldi antenna is shown in **Figure 5** at different resonant frequencies 0.4, 0.5, 0.75, 1.5, 1.75, and 2 GHz. The current distribution of the proposed antenna is studied to verify the operation of the proposed Vivaldi antenna. The exponential edge is responsible for the fundamental resonant frequency of the proposed antenna at 1.75 GHz as shown in **Figure 5**. The semicircular slots are etched to create the frequencies at 0.5 and 0.75 GHz. By adding stubs with different lengths, they are affecting the resonance from 1 to 2 GHz. The highest magnitude of current (red) is related to the corresponding radiating element. The simulated antenna gain of single and dual substrate is shown in **Figure 6(a)**. The gain gives better performance by using dual substrate layer, and it increases from 8 to 17 dBi in average over the operating band from 0.2 to 2 GHz. The average radiation efficiency is around 80% over the operating band for dual substrate, while its value is 50% for single substrate as shown in **Figure 6(b)**.

To validate the simulated results of the proposed antennas with single and dual substrate, they are fabricated by using printed circuit board (photolithographic)

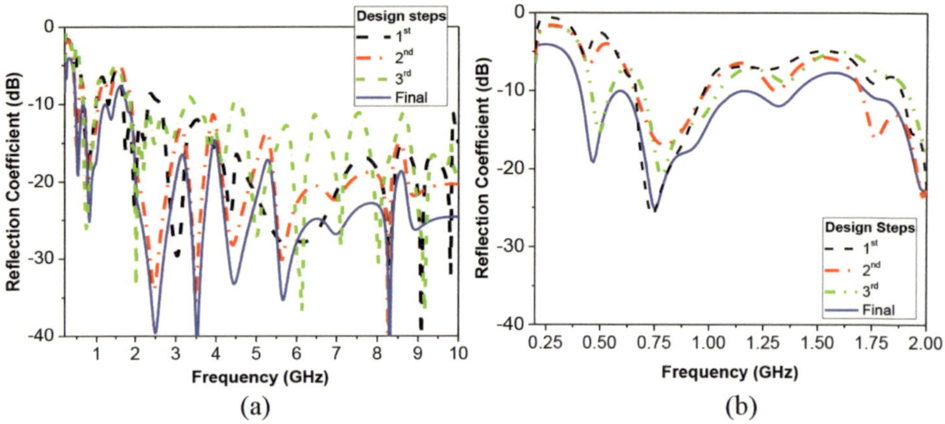

Figure 4.
$|S_{11}|$ versus frequency for the design steps of the proposed antennas (a) whole operating band and (b) zoom on low operating frequency.

Lsub	*Wsub*	*2R*	*Wslot*	*W1*	*W2*	*W3*	
13	12	8	11.5	2.1	1.2	0.6	
W4	*L1*	*L2*	*L3*	*L4*	*Dslot*	*Lslot*	*gslot*
3	3	4	4	3	2.4	4	0.6
LVfe	*LHfe*	*Wfe*	*Led*	*Sd*	*P*	*2r*	
4.5	3.5	1.1	1.5	1.9	0.9	0.6	

Table 2.
Dimensions of the proposed antenna (all dimensions in cm).

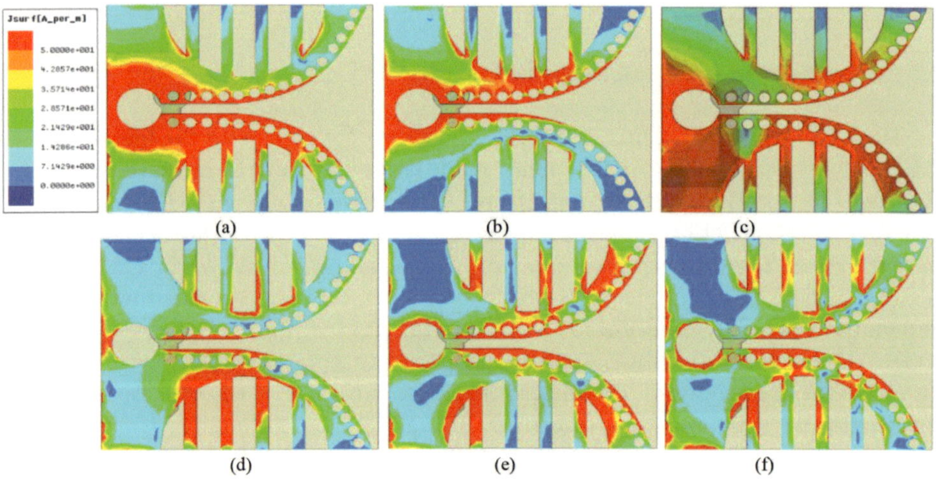

Figure 5.
(a–f) Surface current densities for proposed Vivaldi antenna at 0.4, 0.5, 0.75, 1.5, 1.75, and 2 GHz, respectively.

technology and measured by using Agilent vector network analyzer technologies "Field Fox" Microwave Analyzer N9918A 26.5 GHz. **Figure 7** shows the photo of the fabricated antenna, feeding line, and Vivaldi antenna with dual substrate. The comparison between the measured and simulated results has been performed for the proposed Vivaldi antenna with single substrate that indicates good agreement.

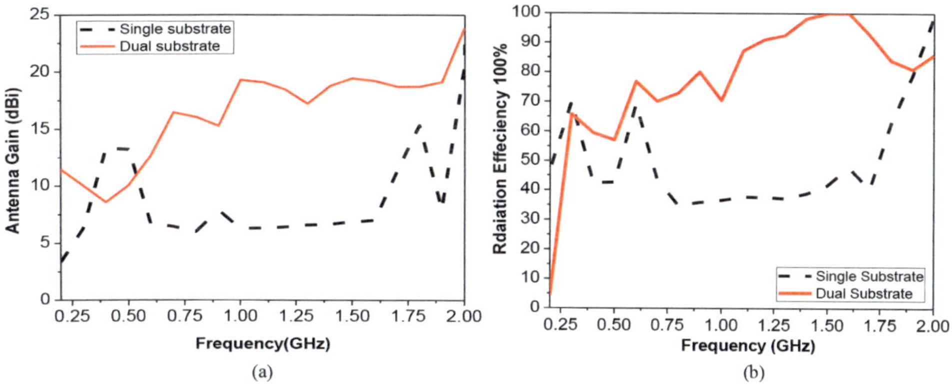

Figure 6.
(a) Vivaldi antenna gain variation versus frequency and (b) radiation efficiency versus frequency of the proposed antenna.

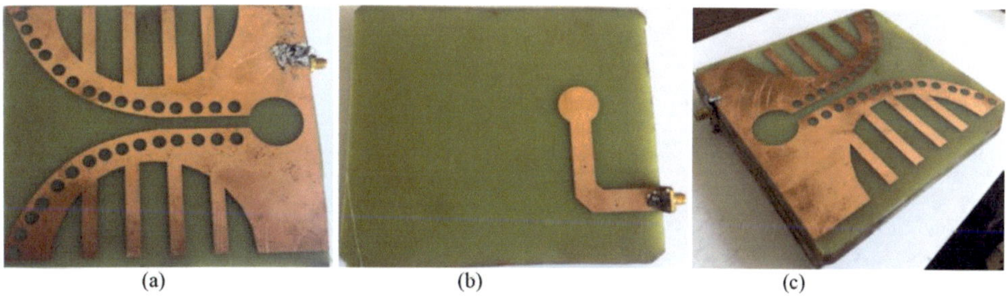

Figure 7.
Photo of the fabricated antenna, (a) top view of one substrate, (b) the feeding line, and (c) 3D of dual substrate.

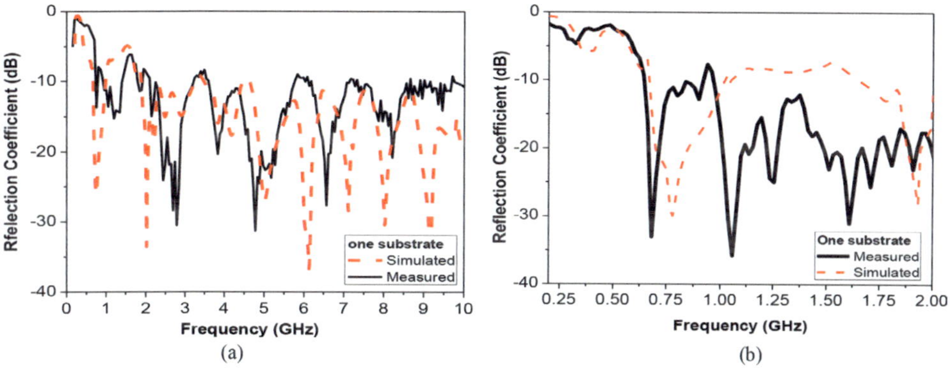

Figure 8.
Reflection coefficient versus frequency for the proposed one substrate layer of Vivaldi antenna.

Figure 8(a) shows the whole range of the operation presenting proposed antenna from 0.2 to 10 GHz, while **Figure 8(b)** shows the zoom-operating range from 0.2 to 2 GHz. The measured reflection coefficient of proposed Vivaldi antenna with dual substrate is shown in **Figure 9**. **Figure 9(a)** shows the whole range from 0.2 to 10 GHz, while **Figure 9(b)** shows the zoom-operating range from 0.2 to 2 GHz. One can notice that there is a slight difference between the measured and simulated results of the reflection coefficient due to soldering of feeding launcher and fabrication tolerances.

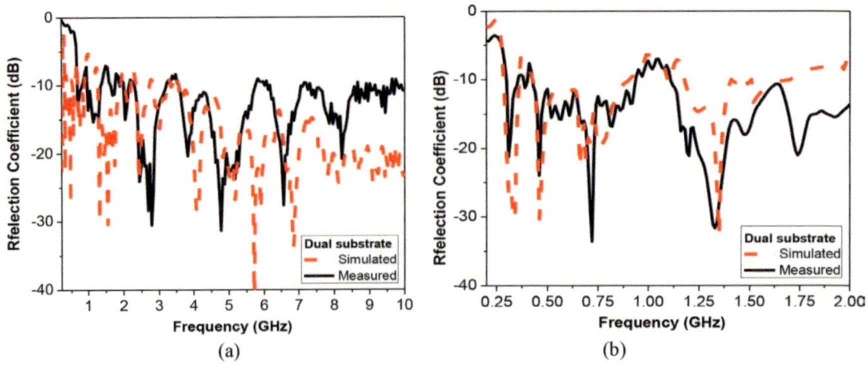

Figure 9.
Reflection coefficient versus frequency for the proposed two substrate layers of Vivaldi antenna.

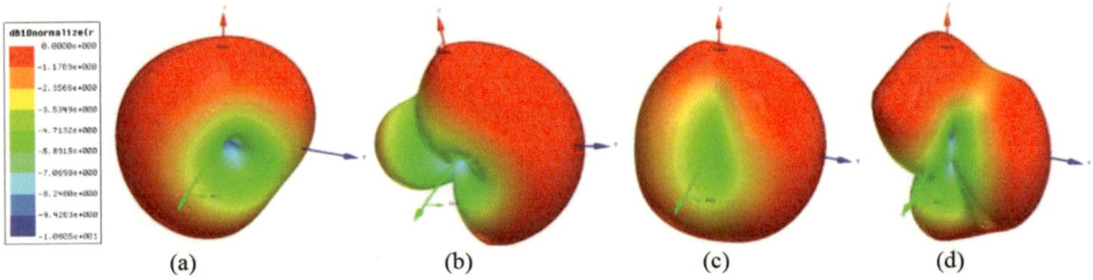

Figure 10.
(a–d) 3D radiation pattern of dual layer substrate at 0.5, 1, 1.5, and 2 GHz, respectively.

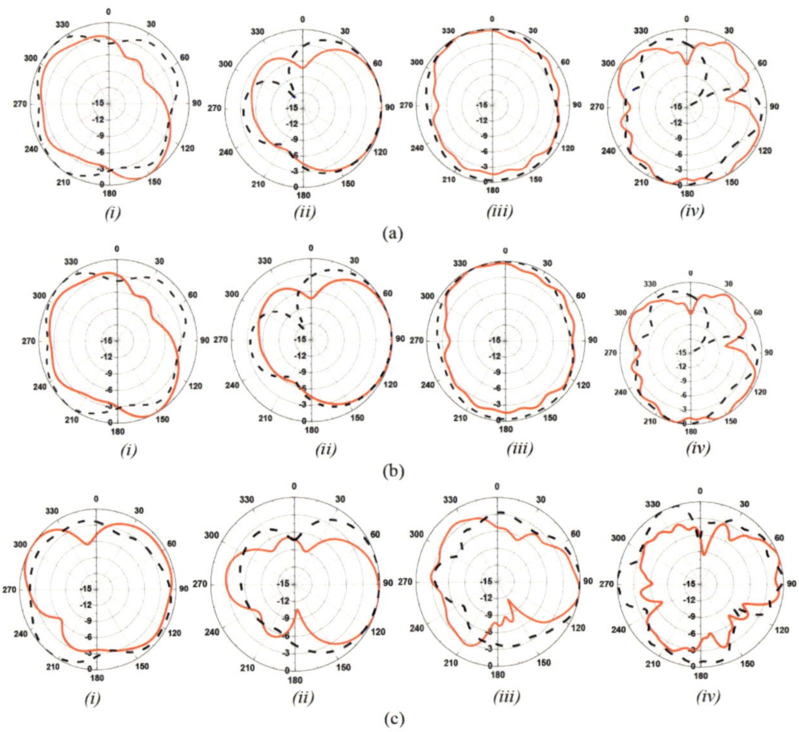

Figure 11.
From (i) to (iv) 2D radiation pattern at 0.5 GHz, 1 GHz, 1.5 GHz, and 2 GHz, respectively, of single and dual substrate (a) Φ°=0°, (b) Φ°=90°, of single and dual substrate, and (c) from (i) to (iv) θ=90°, (- - - one substrate antenna, and ——— dual substrate antenna).

Figure 10 shows the three-dimensional radiation patterns of the proposed dual substrate Vivaldi antenna at different operating frequencies within the operating band at 0.5, 1, 1.5, and 2 GHz. The radiation patterns correspond to the axes shown in **Figure 1**. In the antenna, the radiator and the ground plane are participating to radiation. End-fire radiation pattern is an important requirement for ultra-wideband GPR application system. At lower frequencies of operation, the pattern resembles a conventional dipole antenna, but at higher end of the UWB spectrum, a few ripples are observed which is due to higher-order modes.

4. Printed quasi-Yagi antenna with size reduction for water detection

Nowadays, quasi-Yagi printed antenna is extensively used in modern radar systems due to some advantages as high directivity, good radiation efficiency, affordable, low profile, and easy fabrication [33-35, 37]. However, the disadvantage of these antennas is narrow bandwidth, which achieves about 10%. So, the microstrip-fed quasi-Yagi antenna was initially introduced in 1991 [38] to improve the bandwidth of planar printed quasi-Yagi antennas, and many designs have been reported in [39]. A quasi-Yagi antenna based on microstrip-to-slot-line transition structure was presented in [40]. Modified wideband microstrip-to-coplanar strip-line (CPS) balun was used in quasi-Yagi antenna designs for increasing the antenna bandwidth [41]. Approximately 48 and 38.3% bandwidths were achieved by using the microstrip-to-coplanar strip-line transition structures in [42] and [43], respectively. However, the antenna bandwidths are still restricted by the delay line used in the balun structures. Coplanar waveguide feeding or ultra-wideband balun was presented to improve the bandwidth in some designs [44]. A broad bandwidth of 44% was obtained in [45]. However, the asymmetric nature of the printed quasi-Yagi antenna deteriorates the unidirectional radiation patterns. An ultra-wide band balun feeding structure in which the balun was realized via holes was used in quasi-Yagi antenna for wideband in [46]. Slot and CPS-fed feeding structures were also used in planar printed quasi-Yagi antenna to increase the bandwidth. The maximum available bandwidth of these techniques is about 55%. To improve the bandwidth of quasi-Yagi antenna by modifying the driver to a tapered driver or bowtie driver, rapid developing technology of remote sensing and radar has led to the ultra-wide band (UWB) electronic systems.

4.1 Antenna design and geometry

Figure 12 shows the geometric structure and parameters of the proposed planar quasi-Yagi antenna. This antenna is printed on commercial thick FR4 substrate and a thickness of 9.5 mm. The antenna consists of a microstrip-line-to-slot-line transition structure, a meandered driver T-shaped dipole and two meandered parasitic strips on top layer. The feeding system is printed on the other substrate side with lengths L_f and S_D with a circular resonator. The circular resonator is used to match the input impedance of the antenna to a 50 Ω feeding line. The dimension of the substrate width and length are 72 × 70 cm^2. For matching the antenna, a λ/4 slot line ended with a circular slot of diameter L_D is used.

The traditional printed quasi-Yagi antenna started to resonate from 90 MHz as shown in **Figure 13(a)**. Folded stub ground plane and driven dipole that is etched in order to reduce the size of the printed quasi-Yagi antenna, as shown in **Figure 13(b)**, are employed. The ground plane width and the driver dipole length are equal. The antenna shown in **Figure 13(b)** has 80 MHz as the lowest frequency. Since the ground plane has reduced size, the bandwidth of the antenna is reduced. A driver meander

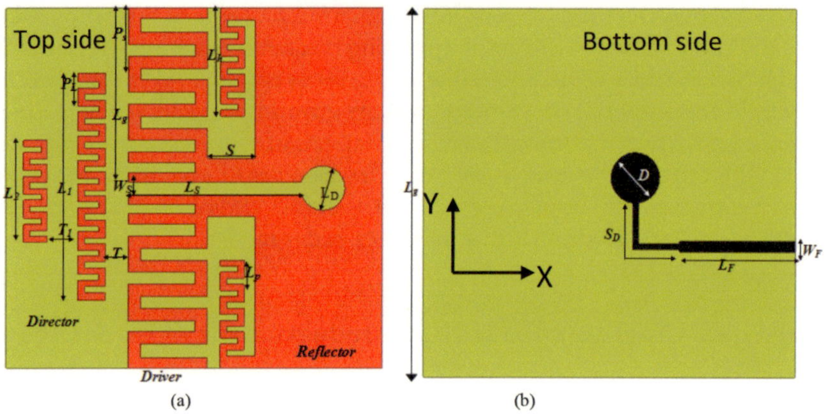

Figure 12.
The quasi-Yagi antenna configuration (a) upper and (b) bottom layer [37].

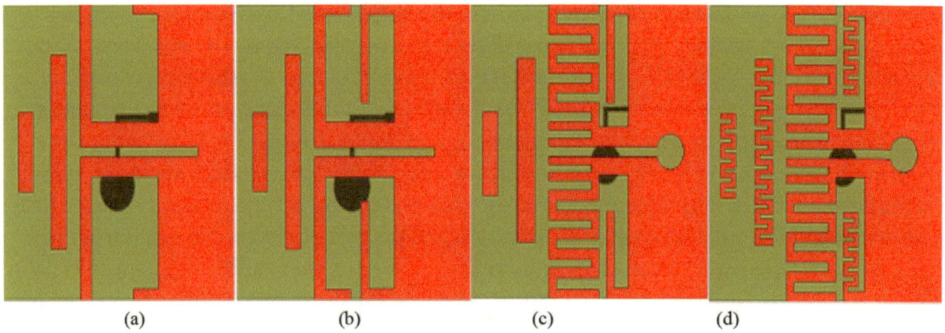

Figure 13.
The compact printed quasi-Yagi antenna design steps [37].

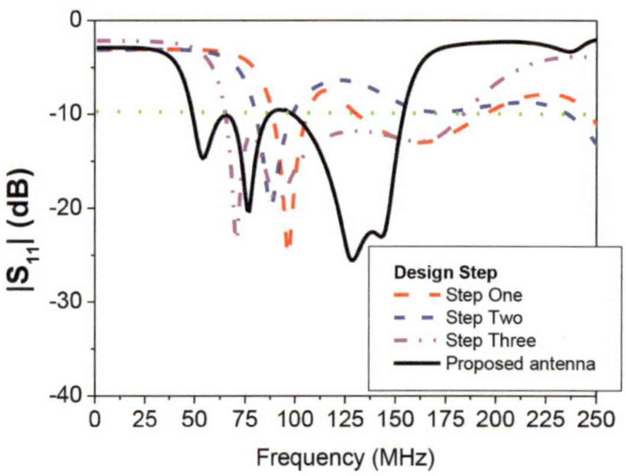

Figure 14.
Simulated $|S_{11}|$ of the design steps as shown in ***Figure 11*** *[37].*

dipole is etched to increase the electrical size of the antenna (**Figure 13(c)**). The antenna started to operate from 62.5 to 185 MHz. Two meandered stubs are symmetrically extended from its ground plane as shown in **Figure 13(d)**. The reflection coefficients |S11| of the antenna design procedure are shown in **Figure 14**.

As shown, the antenna consists of a circular balun feeding which takes the form of a curved microstrip line step transition, in addition to a printed dipole, a ground plane, and two parasitic strips. The larger dipole is located at a distance S away from the ground plane which has a greater length than the larger dipole itself, so it can act

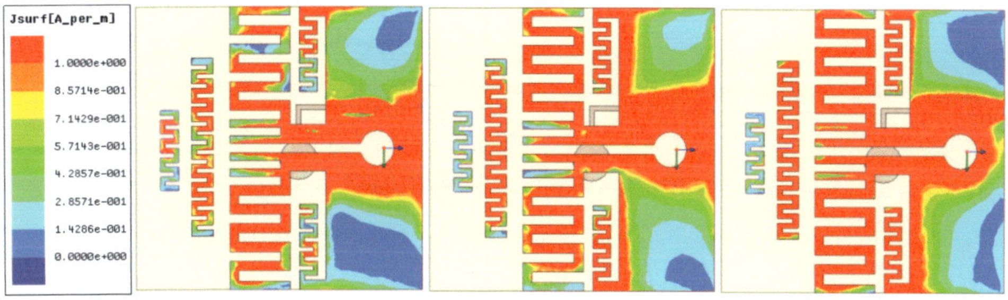

(a) (b) (c)

Figure 15.
The surface current distribution of the printed quasi-Yagi antenna at (a) 50, (b) 100, and 150 MHz [37].

as a reflector. On the second size of the substrate, the two dipoles are located, their line length is 2.6 cm, and the two parasitic spacing was optimized using the reading-made software package HFSS ver.14 in order to improve the antenna performance which means that it has a wide bandwidth, stable radiation, moderate gain, and high front-to-back ratio.

In quasi-Yagi antenna design, metallic strip is always used as a director. To improve the directivity and impedance matching in the high-frequency band, the two metallic strips are used. The simulated surface current distributions of the proposed antenna at 50, 100, and 150 MHz are shown in **Figure 15**. The directors have weak surface current as shown in **Figure 15**, while the two parasitic strips have a large magnitude of surface current at 50 MHz. the largest value of surface current at resonant 100 MHz takes place at larger parasitic strip and the driven dipole. At 150 MHz, the current concentrates on the meandered driven dipole. Compared to **Figure 15**, the surface currents on the metallic strips are enhanced, which means that the effects of the parasitic strips as directors improved the antenna performance at the high-frequency band.

4.2 Antenna parameter study

The antenna structure was optimized to operate at 100 MHz center frequency. The reflection coefficient against frequency for different values of L_1 is shown in **Figure 16(a)**. As the value of L_1 is increased (in steps from 40 to 55 cm), the resonance frequency decreased. At L_1 = 50 cm, the antenna provides the largest beamwidth. One can say that the resonant frequency of the lower band is mainly determined by the length of the larger strip. The dipole in this antenna acts as director of quasi-Yagi antenna. Also, simulation was done to see the effect of L_2 on the antenna bandwidth. **Figure 16(b)** shows S_{11} against frequency for different values of L_2, which shows that as L_2 increased from 20 to 35 cm, the highest frequency almost did not change, while the lower resonance (50 MHz) slightly changed. When the ground plane circular slot L_D was changed from 4 to 5.5 cm, impedance matching was affected, while the lower and higher frequency did not change (**Figure 17(a)**). One can conclude that the diameter of the slot affects the feeding impedance. The effect of the feeding length L_f was studied as shown in **Figure 17(b)**. L_f was varied from 15 to 16.5 cm which caused noticeable changes in the operating frequency band and the value of the reflection coefficient. The length L_f highly affects the impedance matching, and the optimized length of the feeding is 16 cm. **Figure 18(a)** shows the reflection coefficient of the antenna as a function of the spacing between the driver and director (T and T_1).

Increments of the spacing decreased the coupling effect between the parasitic element and the dipole induced significant changes in the reflection coefficient in the operating band region from 50 to 150 MHz, but negligible changes occur in the bandwidth.

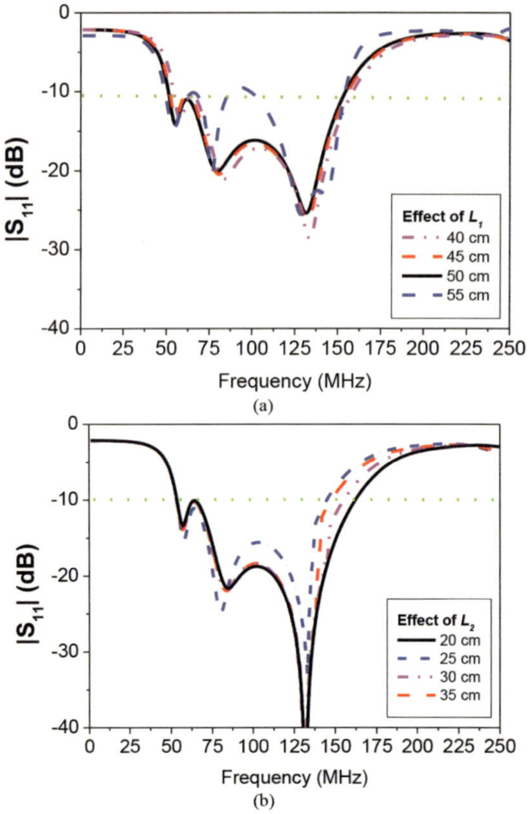

Figure 16.
Effect of the length (a) L_1 and (b) L_2 on the simulated reflection coefficient [37].

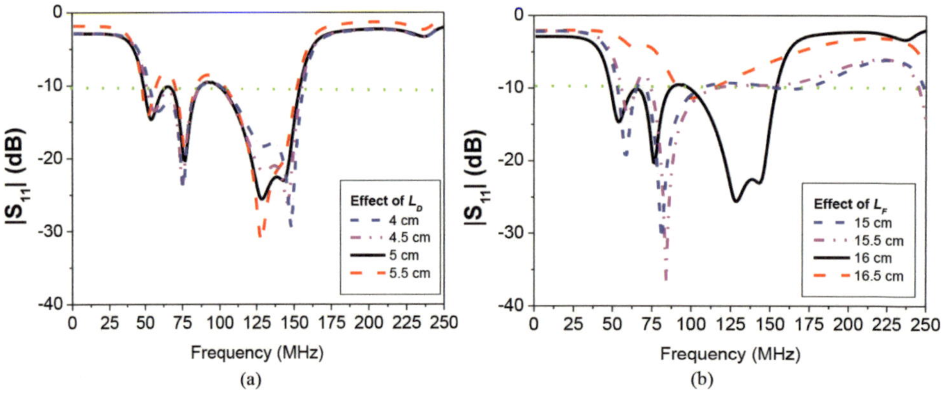

Figure 17.
Effect of the length (a) L_D and (b) L_f on the simulated reflection coefficient [37].

In the high-frequency region of 100 MHz band, the antenna performance is affected mainly by the spacing between the driver and the director. The quasi-Yagi antenna and the T-dipole were designed to operate at 50 and 150 MHz, which shows that a suitable choice of this spacing is very important for the wideband operation of the proposed antenna.

The reflection coefficient S_{11} against frequency is shown in **Figure 19** for different values of the parameter D (balun circle) and S_D (length from the feeding). The lower resonance and impedance matching are varied as D increases from 4 to

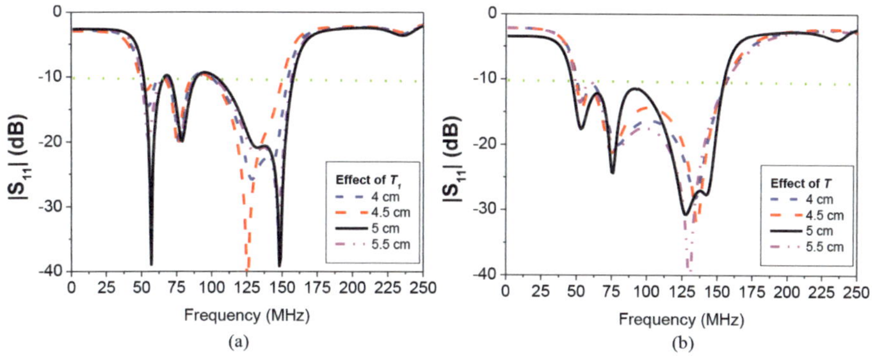

Figure 18.
Effect of the length (a) T_1 and (b) Ton the simulated reflection coefficient.

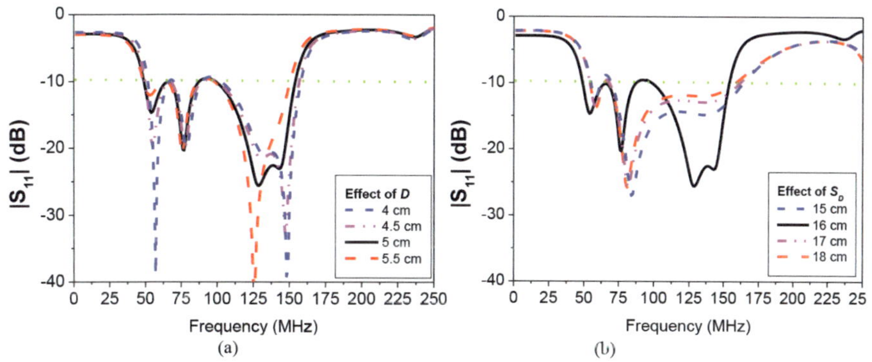

Figure 19.
Effect of the length (a) D and (b) S_D on the simulated reflection coefficient.

5.5 cm. As the value of S_D is changed from 15 to 18 cm, the behavior of S_{11} is changed significantly, so the choice of S_D is very important for the operation of the antenna.

4.3 Ground-penetrating radar antenna system

FMCW GPR system is used for the detection of underground water in the frequency range from 50 to 150 MHz. **Figure 20(a)** shows the radar system which requires a high-gain antenna to obtain acceptable scanning resolution. We used laboratory measurement and EM simulation in order to investigate the electrical and physical properties of the sand and fresh water. The simulated parameters depend on the Debye dispersive model inherent in HFSS software package. **Figure 20(b)** shows the study of the ground effect on the radiation characteristics of the antenna S11 projection. The distance *K* between the antenna and the ground surface was increased from up to 100 cm. The volume of the sand layer was 300 × 200 × 200 cm^3. **Figure 21** shows the reflection with and without the sand layer. It was found that in order to keep S_{11} very close from the case of free space, *K* should not be less than 50 cm. **Figure 22** shows both the gain and radiation efficiency. It is clear that the gain was increased by about 1.5 dBi compared to the case of free space, which may be attributed to the increase in directivity at certain frequencies, while it remains unchanged in other frequencies. The antenna radiation efficiency, as indicated from **Figure 22**, is reduced. The three-dimension radiation pattern of the proposed antenna in both cases is also studied at three different resonant frequencies 50, 100, and 150 MHz, respectively, as shown in **Table 3**, while the 2D radiation patterns at xy-plane ($E\Phi$, $E\theta$) at $\Phi = 0^\circ$ and yz-plane

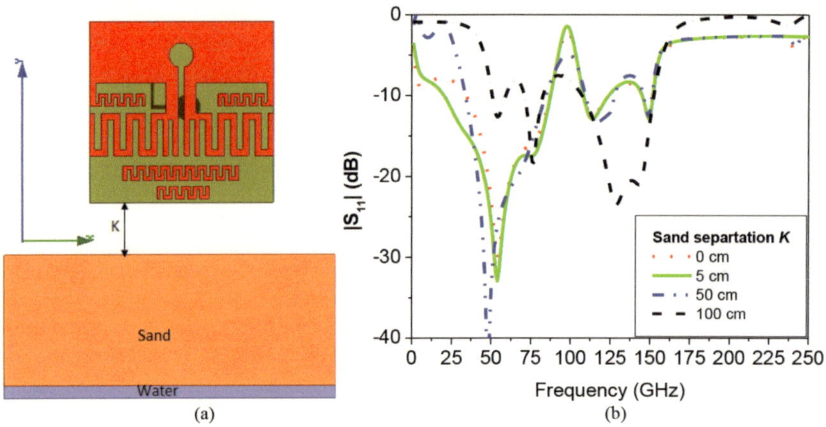

Figure 20.
(a) The GPR antenna system for water detection and (b) the effect of K on proposed antenna reflection coefficient [37].

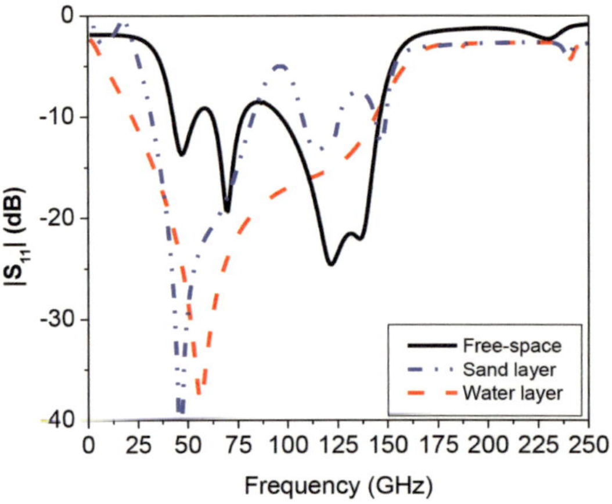

Figure 21.
$|S_{11}|$of the receiver antenna in different cases at K = 50 cm.

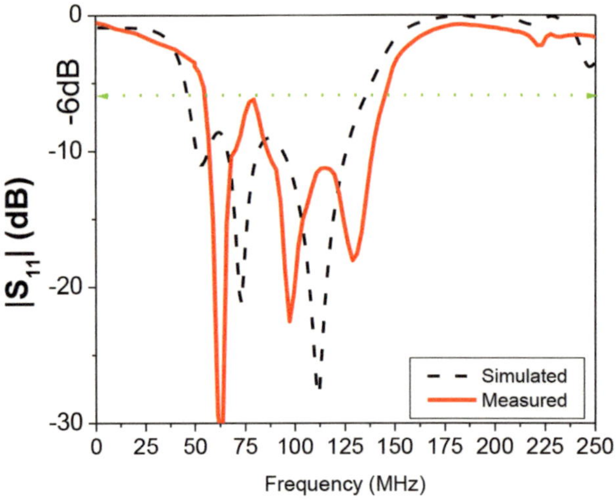

Figure 22.
$|S_{11}|$ comparison between measured and simulated reflection coefficient of the proposed antenna [37].

Free space	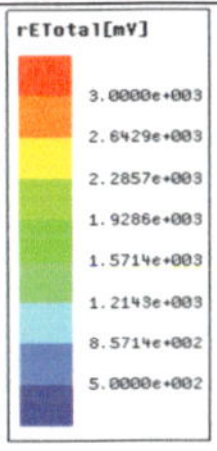	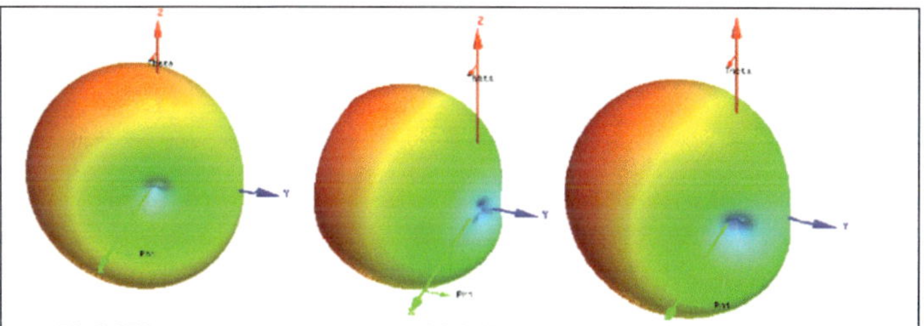 50 MHz 100 MHz 150 MHz
Sand layer	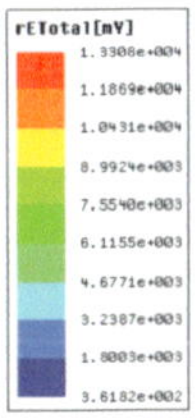	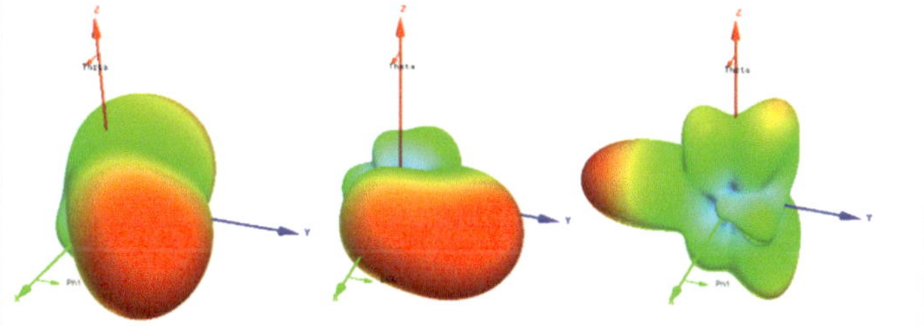 50 MHz 100 MHz 150 MHz

Table 3.
The 3D radiation pattern at different resonant frequencies with and without sand layer

Fre. (MHz)	XY	YZ
50	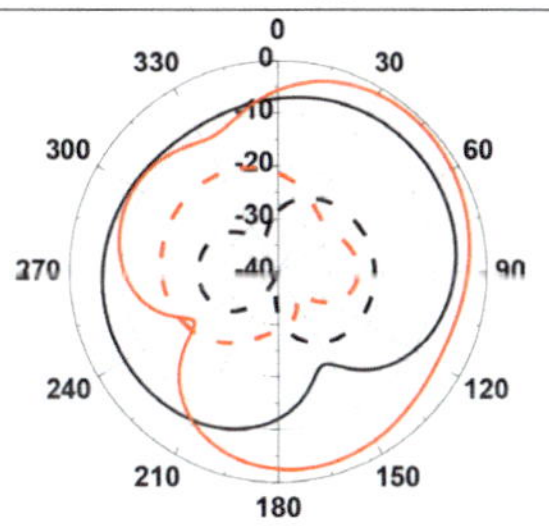	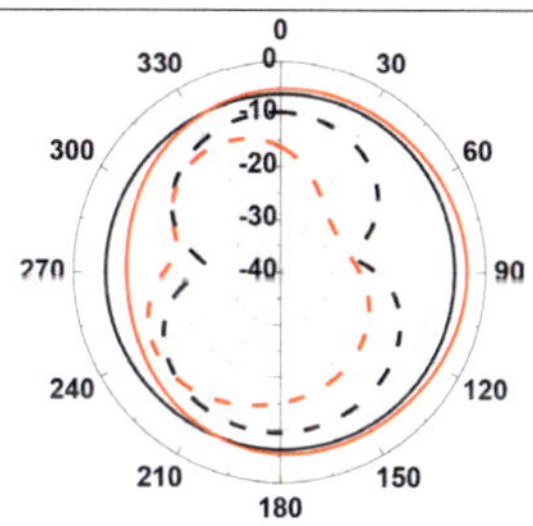
100	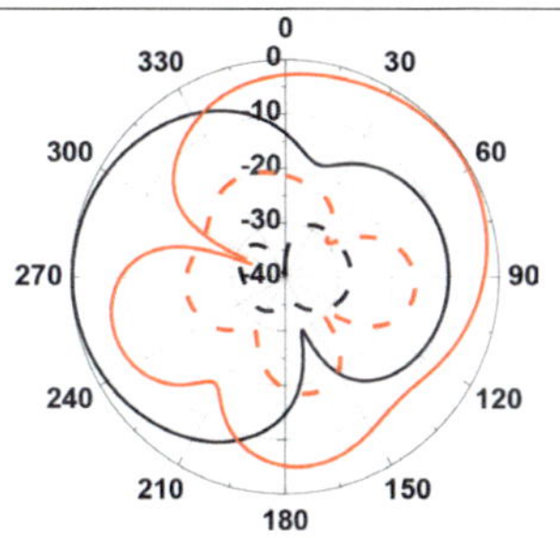	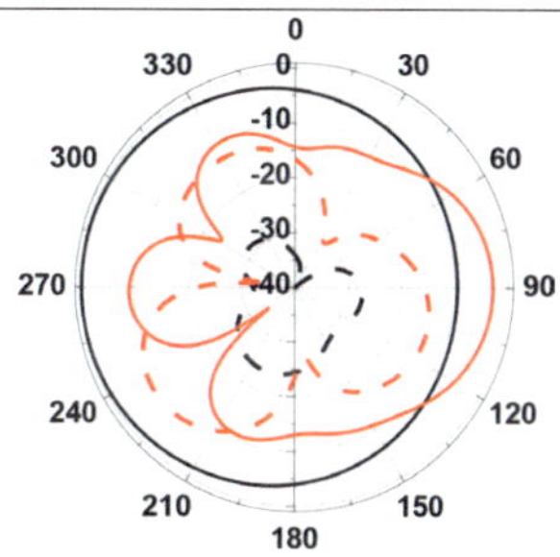
150	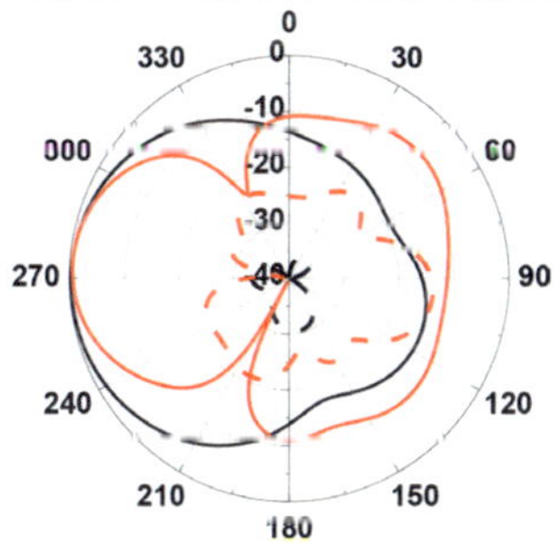	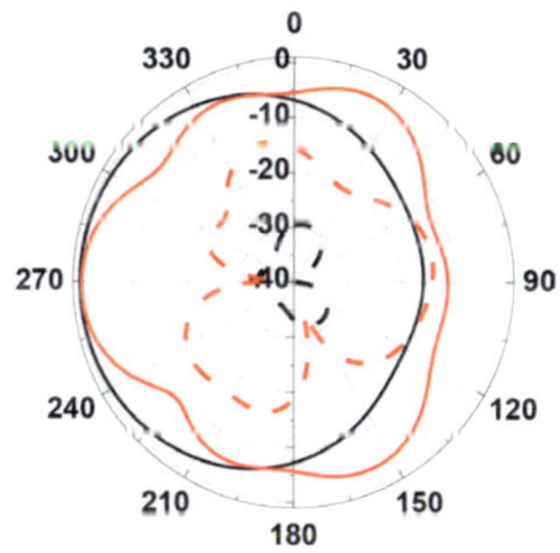

Table 4.
The 2D EΦ and Eθ (phi = 90° and theta = 90°) at frequencies 50, 100, and 150 MHz with and without sand layer, black; without sand, red; with sand layer, solid line (EΦ) and dash line (Eθ), respectively

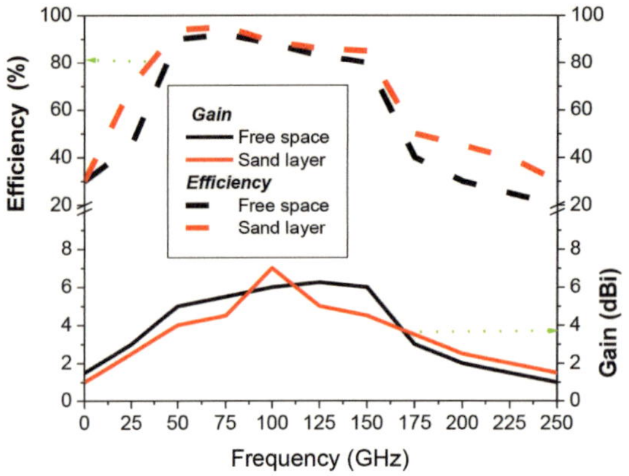

Figure 23.
The antenna gain and efficiency with and without sand layer [37].

($E\Phi$, $E\theta$) at $\theta = 90^{\circ}$ of the original design are plotted in **Table 4** at the same frequencies. The two-dimensional radiation patterns at 50, 100, and 150 MHz with and without sand are platen in **Table 4**. A slight change took place in the E-plane and H-plane radiation patterns keeping good directivity within the whole frequency band. The bock lobes are below −5 dB. Deterioration takes place in the H-plan (YZ plane) as the frequency increases especially in the high-frequency bands (e.g., 500 MHz).

4.4 Fabrication and measurements

A photolithographic technique on FR4 substrate with 100 μm copper thickness was used to realize the proposed antenna. A 50 Ω SMA launcher was used as a transition between the microstrip antenna and the coaxial line, which was not included in the simulation process. Rohde and Schwarz ZVA67 VNA was used to measure S11. The antenna gain and radiation efficiency are also tested as shown in **Figure 23** in both cases namely free space and in the presence of sand layer. The antenna gain is also tested as shown in **Figure 23** in both cases namely free space and in the presence of sand layer. The proposed antenna performance is investigated with and without sand layer as shown in **Figure 20(a)**. The sufficient distance that keeps the antenna reflection coefficient near from free space is almost about 50 cm as shown in **Figure 20(b)**. **Figure 23** shows that the antenna radiation efficiency is increased at present of sand layer by 5%. While the antenna gain is reduced by about 1.5 dBi in the present of sand layer. The gain in the present of sand layer increases at certain resonant frequency due to increase in directivity and has value less than free space on the other resonant frequencies. The optimized antenna reflection coefficient is measured, and there is a good agreement with the simulated results. The measured bandwidth extends from 56 to 150 MHz for the −6 dB reflection coefficient, while the simulated bandwidth extends from 45 to 140 MHz. This bandwidth completely covers the specification for operation.

The slight difference between the measured and simulated reflection coefficient could be from misalignment between curved microstrip line and the circular slot of the balun and effect of the SMA connector, in addition to some fabrication tolerances.

5. Conclusion

The importance of aquifer water increased these days after decreasing the fresh water. A precise approach for the detection of buried nonmetallic objects is ground-penetrating radar (GPR). It should be importance of aquifer water increased these days after decreasing the freshwater. A precise approach for the detection of buried nonmetallic objects is ground penetrating radar (GPR). A novel, miniaturized, and low-cost antennas, which compromise between depth and resolution for different applications, are highly needed. This requires hundreds of conducting research activities in this area. However, although GPR has shown some promising results, the complexity of the problem requires certain challenges on the operation of GPR systems. One of the most serious hardware components for the performance of ground-penetrating radar (GPR) is the antenna system. First, a new compact Vivaldi antenna is designed to achieve ultra-wideband extending from 0.4 to 10 GHz and high average gain of about 17 dBi with average radiation efficiency of about 80%. The proposed antenna consists of dual exponential Vivaldi shapes on both dielectric substrates with the same feeding network. The GPR results in frequency range extending from 0.4 to 2 GHz of the water detection have also been investigated under certain system parameters. Second, ultra-wideband printed quasi-Yagi antenna has been introduced. The proposed antenna has the advantage of simple single feed structure which consists of a circulator balun with a microstrip line and a circulator slot. The antenna bandwidth extends from 47 to 150 MHz based on −6 dB reflection coefficient. The gains at the upper and lower bands are 3.5 and 6.5 dBi. It should be noted that the proposed antenna has the advantage of simple structure and stable end-fire radiation pattern, which makes it widely used in ground-penetrating radar. The average simulated gain in this band is about 5 dBi. At the same time, this antenna possesses low profile and miniaturization characteristics. Furthermore, the wide beamwidth and high gain can promise the antenna to be widely applied in many communication systems.

Author details

Dalia N. Elsheakh[1,2]* and Esmat A. Abdallah[2]

1 Hawaii Center for Advanced Communication (HCAC), Hawaii University, Honolulu, Hawaii, USA

2 Department of Microstrip, Electronics Research Institute, Giza, Egypt

*Address all correspondence to: dalia8@hawii.edu

References

[1] Federal Communications Committee. First Report and Order, Revision of Part 15 Commission's Rule Regarding Ultra Wideband Transmission Systems. FCC 02-48; 2002

[2] Dastranj A. Wideband antipodal Vivaldi antenna with enhanced radiation pattern. IET Microwaves, Antennas and Propagation. 2015;**9**(15):1755-1760

[3] Zheng Y, Guo C, Lin X, Yang X. Vivaldi antenna design in millimeter wave band with ultra wide bandwidth and high gain. In: IEEE MTT-S Workshop Series on Advanced Materials and Processes of RF and THz Applications, Chengdu, China. 2016. pp. 1-3

[4] Pandey GK, Verma H, Meshram MK. Compact antipodal Vivaldi antenna for UWB applications. Electronics Letters. 2015;**51**(4):308-310

[5] Elsheakh DN, Elsadek H, Abdallah EA. Microwave test. In: Noninvasive Electromagnetic Biological Microwave Testing. Rijeka . ISBN:978-953-51-2867-0: InTech; 2017

[6] Elsheakh DN, Eltresy NA, Abdallah EA. Ultra wide bandwidth high gain Vivaldi antenna for wireless communications. In: Progress in Electromagnetics Research Letter (PIERL). Vol. 69. 2017. pp. 105-111

[7] Kalbhor G, Vyas M, Patil BP. Antenna design for ground penetrating radar system. International Journal of Advanced Electrical and Electronics Engineering (IJAEEE). 2013;**2**(6):15

[8] Karamzadeh S, Kiliç OF, Hepbiçer AS, Demirbaş F. Bow tie antenna design for GPR applications. International Journal of Electronics, Mechanical and mechatronics Engineering. 2016;**6**(2):1187-1194

[9] Thaysen J, Jakobsen KB, Lenler-Eriksen H-R. Wideband cavity backed spiral antenna for stepped frequency ground penetrating radar. In: IEEE Antennas and Propagation Society International Symposium, Washington, DC, USA. 2005

[10] Morrow IL, Persijn J, van Genderen P. Rolled edge ultra-wideband dipole antenna for GPR application. IEEE AP-S International Symposium (Digest). 2002;**3**:484-487

[11] Malherbe J, Barnes N. Tem horn antenna with an elliptic profile. Microwave and Optical Technology Letters. July 2007;**49**:1548-1551

[12] Shao J, Fang G, Ji Y, Tan K, Yin H. A novel compact tapered-slot antenna for GPR applications. IEEE Antennas and Wireless Propagation Letters. 2013;**12**:972-975

[13] Zhang F, Fang G-Y, Ji Y-C, Ju H-J, Shao J-J. A novel compact double exponentially tapered slot antenna (DETSA) for GPR applications. IEEE Antennas and Wireless Propagation Letters. 2011;**10**:195-198

[14] Elsheakh DN, Abdallah EA. Compact shape of Vivaldi antenna for water detection by using ground penetrating radar (GPR). Microwave and Optical Technology Letters. 2014;**56**(8)

[15] Elsheakh DM, Abdallah EA. Novel shapes of Vivaldi antenna for ground penetrating radar (GPR). In: 7th European Conference on Antenna and Propagation (EuCAP), Gothenburg. 2013. pp. 2886-2889

[16] Atteia GE, Shaalan AA, Hussein KFA. Wideband partially-covered bowtie antenna for ground-penetrating radars. Progress in Electromagnetics Research. 2007;**71**:211-226

[17] Diamanti N, Redman D, Giannopoulos A. A study of GPR vertical crack responses in pavement using field data and numerical modelling. In: 13th Int. Conf. Ground Penetrating Radar (GPR). Lecce, Italy; 2010

[18] Jol HM. Ground Penetrating Radar Theory and Applications. 1st ed. Amsterdam: Elsevier Science; 2009

[19] Carlotto MJ. Detecting buried mines in ground penetrating radar using a Hough transform approach. In: Battlespace Digitization and Network-Centric Warfare II. Orlando; 2002. pp. 251-261

[20] Daniels DJ. Ground Penetrating Radar. 2nd ed. IET Radar, Sonar, Navigation and Avionics Series 152007

[21] Hoole JG. Implementation of a Low-Cost FM-CW Radar. M.S. in Electronic Engineering at the University of Stellenbosch South Africa; 2008

[22] Amiri A. Multi-band and dual polarized ultra-wide band horn antenna for landmine detection using ground penetrating radar technique. [Thesis submitted for the degree of Doctor of Philosophy]. University College London; 2015

[23] Liu Y, Wang M, Cai Q. The target detection for GPR images based on curve fitting. In: 3rd International Congress on Image and Signal Processing. Yantai, China; 2010. pp. 2876-2879

[24] Marshall Legacy Institute. Dogs for the Detection of Landmines. United states Institute of Peace; 2005

[25] Bonham-Carter GF. Geographic Information Systems for Geoscientists: Modeling with GIS. New York: Pergamon Press; 1994

[26] Bressan MA, Anjos CD. Techniques of remote sensing applied to the environmental analysis of part of an aquifer located in the São José Dos Campos region SP, Brazil Environmental Monitoring and Assessment. 2003;**84**:99-109

[27] Bressan MA, Anjos CD. Techniques of remote sensing applied to the environmental analysis of part of an aquifer located in the São José Dos Campos region SP, Brazil. Environmental Monitoring and Assessment. 2003;**84**:99-109

[28] Carranza EJM. "Geochemical Anomaly and Mineral Prospectivity Mapping in GIS" Handbook of Exploration and Environmental Geochemistry. Amsterdam: Elsevier; 2008

[29] Carrière SD, Chalikakis K, Sénéchal G, Danquigny C, Emblanch C. Combining electrical resistivity tomography and ground penetrating radar to study geological structuring of karst unsaturated zone. Journal of Applied Geophysics. 2013;**94**:31-41

[30] Pettinelli E, Burghignoli P, Pisani A, Ticconi F, Galli A, Vannaroni G, et al. Electromagnetic propagation of GPR signals in Martian subsurface scenarios including material losses and scattering. IEEE Transactions on Geoscience and Remote Sensing. 2007;**45**:1271-1281

[31] Shaik AG, Siddiqui MAA. Dielectric properties of calcified tissue-the bone. International Journal of Science, Environment and Technology. 2013;**2**(6):1412-1420

[32] Liang X. Analysis and research on several wide band antenna and fractal antenna. Dissertation for the doctor degree of Xi'an Jiaotong University, China; 1997

[33] Rahim MKA, Karim MNA, Masri T, Asrokin A. Comparison between straight and U shape of ultrawide band microstrip antenna using log periodic technique. In: IEEE International

Conference on Ultra-Wideband, 2007. Vol. 2. ICUWB; 2007. pp. 696-699

[34] Kalbhor G, Vyas M, Patil BP. Antenna design for ground penetrating radar system. International Journal of Advanced Electrical and Electronics Engineering(IJAEEE). 2013;**2**(6):1-5

[35] DeJean GR, Thai TT, Nikolaou S, Tentzeris MM. Design and analysis of microstrip Bi-Yagi and quad-Yagi antenna arrays for WLAN applications. IEEE Antennas and Wireless Propagation Letters. 2007;**6**:244-248

[36] Elsheakh D. Electromagnetic band gap (EBG) structure for microstrip antenna systems (analysis and design)" [Thesis Submitted for the Degree of Doctor of Philosophy]. Ain Shams University faculty of Engineering, Electrical and Communication Dept.; 2015

[37] Elsheakh DN, Abdallah EA. Ultra wide band planar printed quasi-Yagi antenna with size reduction for water detection in the Egyptian Desert. Microwave and Optical Technology Letters. 2015;**57**(1):226-233

[38] Huang J, Demission AC. Microstrip Yagi array antenna for mobile satellite vehicle application. IEEE Transactions on Antennas and Propagation. 1991;**39**(7):1024-1030

[39] Han K, Park Y, Choo H, Park I. Broadband CPS-fed Yagi-Uda antenna. Electronics Letters. 2009;**45**(24):1207-1209

[40] Kan HK, Waterhouse RB, Abbosh AM, Bialkowski ME. Simple broadband planar CPW-fed quasi-Yagi antenna. IEEE Antennas and Wireless Propagation Letters. 2007;**6**:18-20

[41] Woo DS, Kim YG, Kim KW, Cho YK. Design of quasi-Yagi antennas using an ultra-wideband balun. Microwave and Optical Technology Letters. 2008;**50**(8):2068-2071

[42] Ta SX, Kim B, Choo H, Park I. Slot-line-fed quasi-Yagi antenna. In: Proc. Int. Symp. on Antennas Propagation and EM Theory, Dec. Guangzhou; 2010. pp. 307-310

[43] Nguyen PT, Abbosh A, Crozier S. Wideband quasi-Yagi antenna with tapered driver. In: Proc. Asia-Pacific Conf. on Antennas and Propagation, Singapore. 2012. pp. 138-139

[44] Jiang K, Guo QG, Huang KM. Design of a wideband quasi-Yagi microstrip antenna with bowtie active elements. In: Proc. Int. Conf. on Microwave and Millimeter Wave Technology (ICMMT). 2010. pp. 1122-1124

[45] Wu J, Zhao Z, Nie Z, Liu Q-H. A broadband unidirectional antenna based on closely spaced loading method. IEEE Transactions on Antennas and Propagation. 2013;**61**:109-116

[46] Matthias J, Ammann M, McEvoy P. UWB Vivaldi antenna based on aspline geometry with frequency band-notch. In: Proc. IEEE Antennas Propag. Soc. Int. Symp. San Diego, CA USA; 2008. pp. 1-4

[47] Ta SX, Kim B, Choo H, Park I. Wideband quasi-Yagi antenna fed by microstrip-to-slot line transition. Microwave and Optical Technology Letters. 2012;**54**(1):150-153

[48] Zanzoni A, Montecchi-Palazzi L, Quondam MX. Mint: A molecular interaction database

Chapter 2

Andean Mountain Groundwater, Drinking Water Sources, and Vulnerability: A Case Study in Central Chile

José Luis Arumí, Enrique Muñoz and Ricardo Oyarzún

Abstract

This chapter presents a study of the Diguillín basin in central Chile where geology is dominated by the Nevados del Chillan volcanic complex. The headwater of the basin has two watersheds: Renegado creek and Alto Diguillín. The hydrogeology was studied using field surveys, streamflow gauging, environmental tracers, and a hydrological model. Surface water balance does not fit for both watersheds because there is a deficit/excess of superficial runoff. Renegado soils are predominantly sands over a basement composed of fractured rock; infiltration of rain and snowmelt predominates over surface runoff, resulting in about 5 m^3/s of depth groundwater that flows to the Diguillín River, discharging in a cluster of springs located 3 km downstream of the surface connection. Therefore, drinking water availability for the communities located at the Renegado watershed is limited to some springs that are located around the valley. There is a significant expansion of second home construction in the area of the Renegado watershed; because of its skiing and hot springs, it is a major tourism center. Due to the extensive use of septic tanks, located above the highly permeable soils that overlie the fractured rock aquifer, there is concern about how water quality may be affected.

Keywords: mountain groundwater system, volcanic geology, water balance, vulnerability, drinking water sources

1. Introduction

Mountain watersheds are rather complex hydrological systems that provide water resources to downstream communities for irrigation, industrial activities, human consumption, and ecosystem sustainability [1]. Considering that 40% of the world population are dependent on mountainous regions for its water supply [2], a proper understanding of the hydrological processes of mountain systems is critical to ensure the sustainable development of mountain communities.

In mountain watersheds, water can be stored and released by a combination of hydrological components that may include glaciers, snowpack, lakes, and groundwater. Of these storage components, the less understood is groundwater, which in many cases is neglected. However, in some areas, groundwater release can be the only available water source for local communities during the dry season.

Worldwide, there is a consensus that in central Chile, climate change will affect the dynamics of glaciers and snowpack, increasing the amount of melting in spring and early summer and reducing the amount of melting in late summer and early autumn, which is low-flow season in Mediterranean climate areas, like central Chile [3–5]. Thus, groundwater storage and liberation will be more important in terms of the resilience of mountain communities to climate variability, especially in mountainous areas where the presence of fractured porous rock systems produces conditions for the maintenance of minimum flow due to the liberation of groundwater [6–9].

In the Andean watersheds of the Central Valley of Chile (33.5°–41.5°S), there is little information about the role that fractured porous rock groundwater systems play in the generation of streamflow, mainly because most research has been focused on snow hydrology, as snowmelt drives the streamflow generation in central Chile during spring and summer [3–5]. However, in south central Chile (36°–41.5°S), where the Andes Mountains are lower than 3000 [masl], snowmelt ends in mid-January and until the beginning of the rainy season in mid-April; streamflow in the rivers depends on base flow generated by groundwater exfiltration. Therefore, understanding groundwater recharge, storage, and release processes becomes critical to manage mountainous hydrological systems and therefore to protect water resources.

On the other hand, it is important to highlight that Chile, during the last 20 years, has experienced an increase in income levels, which produced a strong demand for second homes, especially in high-demand tourism areas, especially in the central area of the country. One of these tourism areas is the Renegado Valley, which is associated with a world-class ski center that also has hot springs, the "Termas de Chillán" complex. At the end of the last century, the valley was part of a large farm that was exploited for forestry and cattle feeding. In the 1990s, the land was divided, and the tourism-related development process began, resulting in a community that stretches along the 30-km mountain valley without any planning for both drinking and wastewater.

This chapter presents the results of research work that had as an initial objective the study of the hydrology of the Renegado Valley, as water availability was identified as a key limitation for the further development of the area. Therefore, the initial research question was why does the Renegado Creek exhibit a permanent shortage of streamflow during the dry season?

The answer to that question is rooted in the hydrogeological characteristics of the Renegado Valley [10–12]. But that answer, which will be presented in this chapter, raised a second question about the vulnerability of the drinking water sources of the communities that are being developed along the mountain valley.

2. Methods and materials

2.1 Study area

The Diguillín River watershed is located in central Chile at latitude 36.9°S and longitude 71.4°W (**Figure 1a**) and drains the southwestern section of the Nevados de Chillán volcanic complex, located in the Andes Mountains (**Figure 1b** and **c**). At the upper part of the watershed, there are two gauging stations that define the two sub-watersheds that are shown in **Figure 1c**: Alto Diguillín (207 km^2) which is controlled by the Diguillín en San Lorenzo (DSL) gauging station and Renegado Valley (127 km^2) which is controlled by the Renegado en Invernada (RI) gauging station.

In the Alto Diguillín sub-watershed, there is a national protected area called Reserva Ñuble and some farms dedicated to forestry and cattle production; in contrast the Renegado Valley, as described before, has been intensively populated for second homes, due to the tourism value associated with the Volcan Chillán Ski Area and the existence of hot springs. Additionally, there is a marked difference between both sub-watersheds when the streamflows are compared. The Renegado Creek exhibits much lower values than the Alto Diguillín (**Figure 2**), more than can be easily explained based on watershed extent. In fact, the Renegado Creek exhibits lower specific flows (flow rate per unit of area), in comparison with those of the neighboring watersheds Alto Diguillín and Chillán (used for comparison in **Figure 3**), even when those rivers exhibit the same East to West orientation and a similar rainfall distribution. That lower specific flow is consistent with the water availability limitation for the development of the community along the valley, which is one of the research questions of this work.

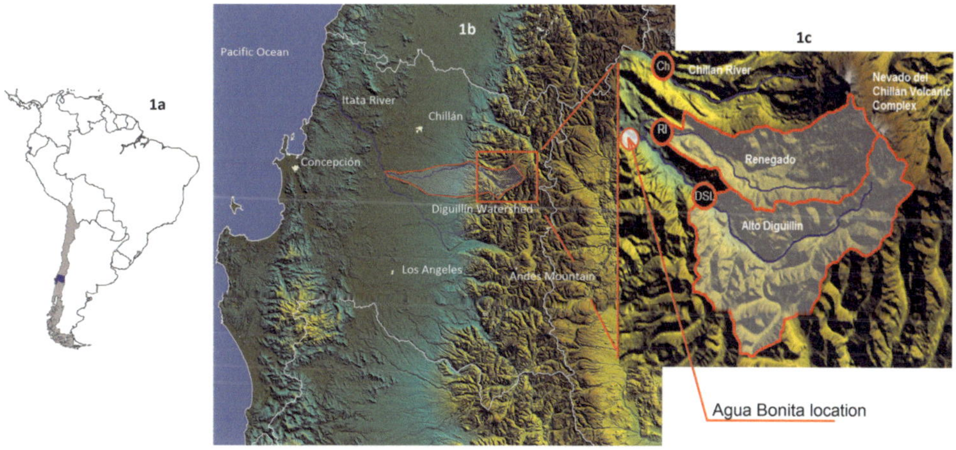

Figure 1.
(a) Diguillín watershed location in South America; (b) the Biobío Region, showing the main cities of Concepción, Chillán, and Los Angeles; (c) location of the Renegado Creek and Alto Diguillín sub-watersheds at the upper section of the Diguillín watershed, the Agua Bonita location where a large cluster of springs flows to the Diguillín River, and also the location of gauging station: "Ch" is Chillán River, "RI" is Renegado en Invernada, and "DSL" is Diguillín en San Lorenzo.

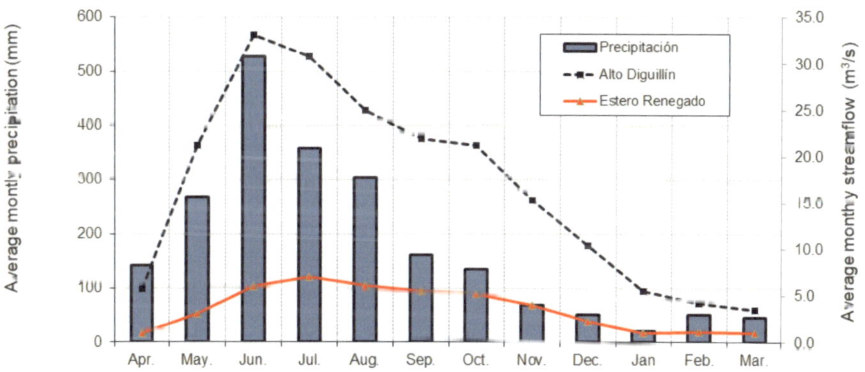

Figure 2.
Average rainfall and streamflow measured at the Diguillín River at San Lorenzo and Renegado Creek in Invernada.

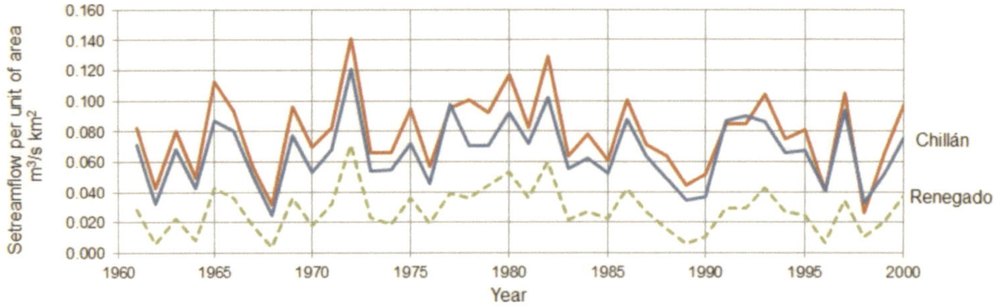

Figure 3.
Comparison of measured monthly specific streamflow for Alto Diguillín and Chillán Rivers with Renegado Creek.

2.2 Field research

Due to the existence of several infrastructure projects that have been proposed for the Diguillín River watershed, previous studies were considered as the base for the initial characterization of the watershed. The study for irrigation planning conducted by the National Commission of Irrigation (CNR) in the Itata River basin, which concludes that the Diguillín River receives flow from groundwater discharge in the middle part of the watershed, which becomes relevant during the low-flow season between January and April [13], was particularly important. Hydrological data for the watershed (streamflow and rainfall) were collected from the database of the Chilean Water Authority (Dirección General de Aguas, DGA).

In addition, to incorporate local knowledge about the Diguillín River, a series of interviews of various stakeholders such as the river authority (Junta de Vigilancia), villagers of every sector, mountaineers, and sport fishermen was carried out in order to determine if the existence of springs that feed the Diguillín River was true.

The available geological information came from two principal publications that describe the geology of the upper part of the Diguillín River watershed [14, 15]. Both references explain the marked influence of the volcanic processes associated with the Nevados de Chillán Complex on the development of this watershed. The geological information [15] includes a geological map at 1:50.000 scale which was digitalized in a raster format and virtually mounted on Google Earth, using Global Mapper software.

As a complementary analysis for the identification of hydrological processes, a hydrogeochemical data analysis was performed. Samples of rain, snow, surface water, and springs collected from the Renegado, Diguillín, and Chillán Rivers during 2012 and 2013 were considered. Samples were chemically analyzed for mayor cations and anions (i.e., Na^+, K^+, Ca^{2+}, Mg^{2+}, Cl^-, HCO_{3-}, SO_4^{2-}) in the Laboratory of Soil and Plants Analysis of the University of Concepción. Additionally, concurrent samples were derived to the Chilean Commission of Nuclear Energy for the environmentally stable isotopes analysis (^{18}O y ^2H). Further description about the technics used can be found in Arumí et al. [10]. Also, samples were analyzed for ^{222}Rn by the Environmental Laboratory of University of La Serena using a Durridge RadH_2O equipment [16].

The analysis of secondary information suggested the existence of a cluster of springs discharging into the Diguillín River, in a gorge located downstream of the confluence of the Renegado Creek and the Diguillín River. This sector was studied in detail by walking surveys, which allowed the identification of a 2-km section of the river with clusters of fractured rock-related springs that discharge to the

Figure 4.
(a) Large spring draining to the Diguillín River in Agua Bonita; (b) water is channeled through multiple small fractured rock springs (b); (c) springs are located at the base of cliffs of a gorge that can be reached only in summer conditions when the river flows are minimal; (d) typical fractured rock profiles observed along the Renegado sub-watershed at the location of the spring cluster.

Diguillín River (**Figure 4**) in a location locally known as "Agua Bonita" (**Figure 1c**). All the springs were located at the bottom of a hundred-meter-high cliff, in a very difficult-to-access area located along a gorge that can be reached only in summer when river flows are minimal.

Because locations were of difficult access, measurements taken in Agua Bonita were only possible at the end of the dry season (March 2012 and 2013). These measurements were carried out using the FlowTracker Acoustic Doppler Velocimeter, from SonTek. Streamflow was measured at the Diguillín River, above and below this 2-km section. It was found that the river flow increases from 2.5 to 7.4 m^3/s; therefore, spring discharge was estimated as 4.9 m^3/s.

2.3 Water balance analysis

The water balance was analyzed through a conceptual model approach to better understand the hydrologic behavior of the Renegado and Alto Diguillín watersheds [12]. The model simulates the rainfall-runoff and snowmelt-runoff processes. The rainfall-runoff component was modeled through a lumped model that considered the watershed as a double storage system: subsurface and groundwater. The snowmelt-runoff model calculates the snowfall based on precipitation above the zero-degree (base temperature at which melting starts) isotherm falling as snow. The melting calculations are performed based on the concept of the degree-day method [17]. Thus, the potential melting is estimated, and then based on the stored snow, the real melting is calculated. The model needs the rainfall and the potential evapotranspiration as inputs, and the output is the total runoff at the watershed outlet, including both subterraneous and direct runoff, the amounts of which are calculated through six calibration parameters, plus two for the input modification (useful in the case of non-representative PM and PET data). Further description about the model, its implementation, and calibration can be found in [12].

The major findings in the water balance of the Renegado-Diguillín system were that the low specific flow condition at the Renegado Valley and the existence of the cluster of springs that flow into the lower Diguillín River suggest that a significant part of the base flow that is produced at the Renegado Valley is transferred through a subterranean connection to the spring cluster [12]. To reproduce such conditions, the Renegado-Diguillín model was modified by adding a groundwater connection, where a percentage of the Renegado base flow was transferred to the Diguillín watershed. This finding is consistent with an indirect estimation of groundwater storage evolution [11] based on recession flow analysis. In that work it was shown that whereas for the Upper Diguillín basin and the period 1961–2010, no increase or decrease trend in groundwater storage was detected; for the Renegado sub-basin, it was possible to observe a statistically significant decreasing trend in subsurface water storage.

This water balance analysis allowed the understanding of the observed condition, i.e., that the Renegado Creek presents lower specific flows than the Alto Diguillín River. By adding a groundwater connection between watersheds, it was possible to better simulate the monthly flows of these two basins. Thus, a main conclusion from these studies was that a groundwater contribution provided from the Renegado watershed to the Diguillín watershed was necessary to adequately reproduce the hydrogeological behavior of the Renegado-Diguillín hydrological system. After the calibration processes, it was possible to estimate that about 77% of the base flow is lost through groundwater seepage from the Renegado watershed. That flow was estimated to be around 4.6 m^3/s, very close and on the same order of magnitude to the 4.9 m^3/s measured at the springs cluster located in the Diguillín River.

3. Results and discussion

3.1 A plausible explanation for the groundwater connection

As hydrological processes should be highly connected with the geological features in fractured rock settings, it is important to give a look into local geological and lithological conditions in order to have a wider insight into the system under study.

The geology of the upper section of the Diguillín watershed is strongly influenced by the volcanic processes associated with the Nevados del Chillán volcanic complex [15, 16]. This volcanic complex is composed of several types of structures created by different processes that have occurred for approximately 650,000 years [15].

The Nevados del Chillán volcanic complex possesses cold and hot springs distributed along its edge [15], from which it may be inferred that the volcanic complex behaves in a form similar to the systems described by [6, 7]. The existence of cold and hot springs indicates the existence of at least two aquifers: a superficial one that receives its recharge by infiltration of rainwater and snowmelt discharging in cold springs and a deeper system, which is recharged from the superficial system and is in contact with the magma chamber, heating the water and producing vapor that feeds the thermal springs (**Figure 5a**).

As well as the thermal springs around the volcanic complex, there is a large cluster of hot and cold springs in the "Valle de Aguas Calientes" (Hot spring valley) where the headwater of the Diguillín River is located (**Figure 5b** and **c**). This cluster is due to local tectonic features related to the formation of valley, like the fault line shown at **Figure 5a**. This geological trait enhances recharge from snowmelt, rainfall, and runoff from adjacent watersheds to the Alto Diguillín sub-basin, explaining why it has more water than surrounding rivers as can be deduced from **Figure 3**.

In relation to the lower specific flow in the Renegado Creek sub-watershed, particular importance is to be given to the formation of the lava units that filled the valley of the Renegado Creek (**Figure 6**). In effect, this valley was formed from a sequence of lava flows. An earlier lava flow called the Pincheira lavas, of the middle Pleistocene, cut along a large glacier forming walls that give the valley its characteristic U shape; at the end of the glacier, the lava flow opened in what is today the locality of Los Lleuques. Later lava flows (Diguillín of the middle Pleistocene) went down the valley until being blocked by the Pincheira lavas, which forced them to turn toward the south, closing the Renegado valley and forcing a connection with the Diguillín River (**Figures 6** and 7).

Two additional lava flows that fill the valley covering the Pincheira lavas are the Atacalco lavas (of the Middle-Upper Pleistocene, which correspond to one or more andesitic lava flows, with a layer thickness of 125 m) and the Democrático Volcano

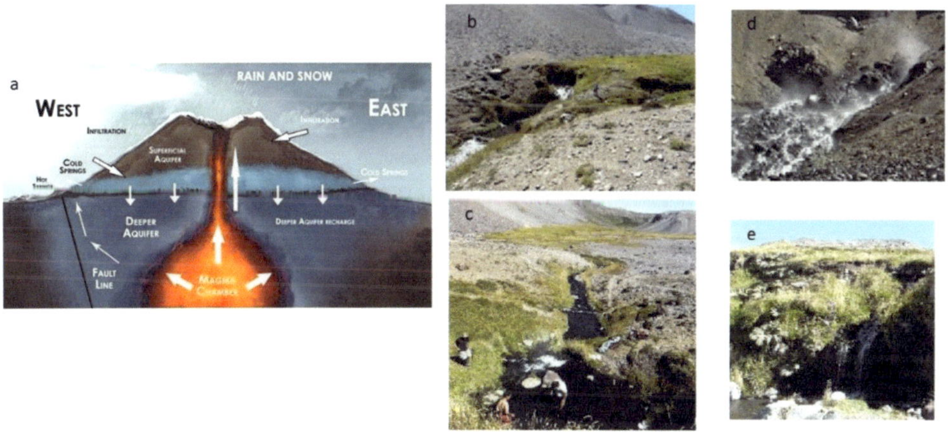

Figure 5.
(a) Scheme of the plausible groundwater system at one volcanic complex; (b) headwaters of the Diguillín River at a thermal spring; (c) Diguillín River; (d) hot water spring; (e) cold water spring.

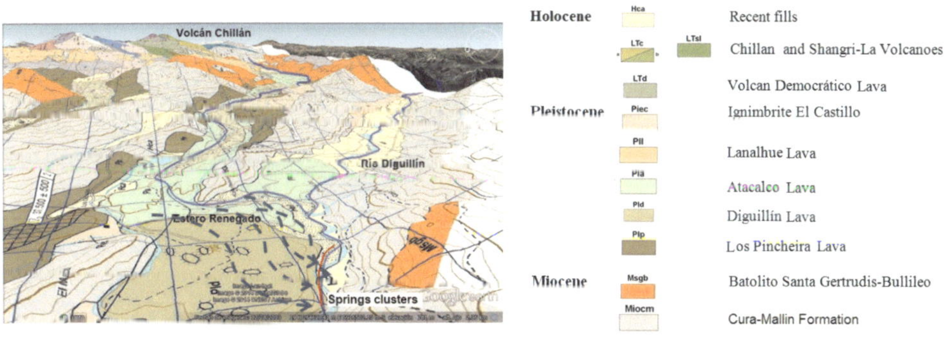

Figure 6.
Geology of the upper section of the Diguillín River adapted from [15] and pasted on Google Earth. The red line at the right corner indicates a river section where fractured rock springs are located; the dashed blue line represents probable groundwater paths.

(a) (b)

(c)

Figure 7.
(a) Pincheira lavas, (b) Atacalco lavas, and (c) Democrático Volcano lavas, Las Trancas.

lavas (LTd) of the Holocene, which are a fundamentally effusive structure of silicious, andesitic to dacitic block lavas (**Figure 7**).

The existence of the Agua Bonita springs is related to the formation of Diguillín lava. It is possible to assume that these lavas entered a postglacial lake which, through the cooling process [15], produced the fracture system which can be observed along the Renegado sub-watershed (**Figure 4d**). The presence of this fractured system causes the groundwater watershed boundary to differ from the surface watershed. In fact, groundwater is moving along paths that were created when the Diguillín lavas filled the valley, and surface water is moving across the watershed created by recent lava flows.

Also, the predominant soils in the upper part of the Renegado valley are sandy soils with high infiltration rates (larger than 200 mm/hr). The existence of these soils on a basement formed by fractured rocks favors groundwater recharge and explains why the Renegado Creek does not have significant superficial runoff. A large amount of rainwater and snowmelt infiltrate into the sandy permeable soil and percolate to the fractured rock system where the water moves through the fractured rock system and discharges in the Diguillín River at the springs described in the previous paragraph.

3.2 Discussion

This case study illustrated how groundwater storage and release can be significant hydrological processes in a mountain watershed where the presence of fractured volcanic rock geology produces the conditions for complex groundwater systems.

In recent years it has been understood that volcanic complexes—such as the Nevados del Chillán complex—produce the conditions necessary for significant mountain groundwater systems. At Mount Fuji in Japan, water can flow vertically through fractures, with water from different aquifer formations mixing, as established using isotopes, major ion chemistry, and multivariate statistical methods [7]. In Mexico, the hydrothermal system of El Chichón volcano was also studied using isotopes [6] and water chemistry, allowing the identification of two aquifers that make up the volcanic structure in a system that is controlled by infiltration from rainfall, water percolation, and heating and production of hydrothermal vapor. In Italy, environmental isotope techniques, hydrogeochemical analysis, and hydraulic data were used to identify recharge areas and trace groundwater flows at Mount Vulture [18].

In a tropical mountain cloud forest catchment located in a volcanic area in Mexico, it was found that rainfall-runoff responses are controlled by rapid vertical rainfall percolation through the high permeable volcanic soils, which recharges the groundwater system, while groundwater storage and discharge modulate the streamflow regime of the catchment [9].

In a mountain watershed without glaciers where volcanic processes are the dominant geological feature, spring discharge plays a major role in streamflow generation [19, 20]. Due to the expansion of second home construction in some mountain valleys, especially those associated with a tourist attraction like ski or hot spring resorts, spring water has become more common as a source of drinking water. However, as the recharge areas are also impacted by housing development, the risk of groundwater pollution increases [21], exacerbating the vulnerability of water quality in mountain groundwater systems [21].

As previously stated, land cover changes in Chile have been driven by an increase in income levels, which has led to significant growth in second home construction in the Renegado watershed area, as it is a major tourism center based on skiing and hot springs. There are now more than 1000 vacation houses and several resorts that have been constructed on more than 5000 small parcels that are available in the area. This explosive increase in construction has taken place without any planning or control, as the area is considered rural land.

The lack of a formal drinking water system has led to a trade in building clandestine catchments that are connected to the slopes by rough plastic pipes. Homeowners pay local people to build illegal water connections, which are unfit to provide drinking water. These connections are not only unhealthy; they also affect the few springs that are located around the valley (**Figure 8a**).

While the situation related to drinking water distribution was referred to in **Figure 8a** as "chaotic," the situation related to wastewater is unknown, but there are reasons to dubiosity. According to Chilean law, disposal of wastewater from small houses located in rural areas should be carried out through the use of septic tanks. With the extensive use of septic tanks, located above the highly permeable soils that overlie the fractured rock aquifer, there is a concern that water quality in the Diguillín River could be impacted by housing and tourism development. Pollutants from the wastewater disposal systems will move through the fractured rock network and discharge into the springs that are used as drinking water sources for the houses and communities that are located down gradient (**Figure 8b**).

In recent years it has been shown that pharmaceuticals and personal care products (PPCP) can be used as indicators of groundwater pollution [22, 23]. A review summarized the use of frequently detected PPCPs, including antibiotics, anti-inflammatories, lipid regulators, carbamazepine, caffeine, and N,N-diethyl-m-toluamide, in groundwater to identify groundwater pollution, analyzing how adsorption to soils and degradation may affect the use of these elements as

Figure 8.
(a) Chaotic drinking water distribution system at the Renegado Valley; (b) scheme of the plausible pollutants' recharge and discharge processes along the Renegado Valley.

groundwater tracers [24]. In groundwater systems such as the Renegado Valley where transit time is expected to be short, adsorption and degradation effects will be less relevant and therefore PPCP would be a good indicator for consideration.

4. Conclusions

In volcanic mountain watersheds, the groundwater system can play an active role in hydrological processes. The groundwater system at the headwater of the

Diguillín River is very active and, at least, has two main subsystems: the existent aquifers located at the volcanic complex itself and the fractured system of the Renegado watershed. Those groundwater systems produce almost all the streamflow of the river at the end of the Chilean summer and early fall.

Each volcano that exists in Chile is a complex aquifer system by itself. There is a lack of knowledge about the groundwater system at the volcanic complexes. The structure of the aquifer systems and the recharge and discharge processes are unknown. Advances in understanding of those processes will allow advantage to be taken from the geothermal potential of the volcanic complexes.

This analysis makes evident the reasons why the Diguillín River has stable minimum flows during the dry season and why the Renegado Creek has a lower specific streamflow. However, those differences were not so obvious 3 years ago, at the start of this research. It is important to emphasize that in practical engineering, the supposition of constant specific streamflow between neighboring watersheds is widely used. Thus, it is important to carefully check this hydrological similarity through an analysis of the climatic and geomorphologic characteristics, soil type, and use. But in watersheds influenced by volcanic systems, it will be necessary to carefully analyze the geological conditions, especially in relation to fractured rock systems.

The highly permeable soil and the fractured rock system in the Renegado subwatershed, where there is significant tourism development and construction of weekend houses, raise questions about the fate of pollutants introduced to the systems by wastewater infiltration from septic tanks. The pathways between pollutant recharge areas and spring discharge are unknown and must be identified and ideally measured in order to improve the sustainable development of the watershed.

Acknowledgements

The authors wish to express their thanks for the support given by the Chilean Scientific Council (Conicyt) through the project FONDECYT 1110298, 11121287, and Conicyt/Fondap/15130015. Also it is important to express our gratitude to the Chilean Dirección General de Aguas and the Comisión Nacional de Energía Nuclear for the support given to some of the works here described.

Author details

José Luis Arumí[1]*, Enrique Muñoz[2] and Ricardo Oyarzún[3]

1 Water Resources Department, College of Agricultural Engineering, CHRIAM Water Center Universidad de Concepción, Chillán, Chile

2 Civil Engineering Department, College of Engineering, Universidad Católica de la Santísima Concepción, Concepción, Chile

3 Mining Engineering Department, College of Engineering, Universidad de La Serena, Ceaza Center, CHRIAM Water Center, La Serena, Chile

*Address all correspondence to: jarumi@udec.cl

References

[1] Viviroli D, Archer D, Buytaert W, Fowler H, Greenwood G, Hamlet A, et al. Climate change and mountain water resources: Overview and recommendations for research, management and politics. Hydrology and Earth System Sciences Discussions. 2010;7:2829-2895. DOI: 10.5194/hessd-7-2829-2010

[2] Buytaert W, Célleri R, De Bievre B, Cisneros F, Wyseure G, Deckers J, et al. Human impact on the hydrology of the Andean Páramos. Earth-Science Reviews. 2006;**79**:53-72. DOI: 10.1016/j.earscirev.2006.06.002

[3] Cortés G, Vargas X, McPhee J. Climatic sensitivity of streamflow timing in the extratropical western Andes Cordillera. Journal of Hydrology. 2011;**405**:93-109. DOI: 10.1016/j.jhydrol.2011.05.013

[4] Vicuña S, Gironás J, Meza F, Cruzat ML, Jelinek M, Bustos E, et al. Exploring possible connections between hydrological extreme events and climate change in central south Chile. Hydrological Sciences Journal. 2013;**58**:1598-1619. DOI: 10.1080/02626667.2013.840380

[5] Ohlanders N, Rodriguez M, McPhee J. Stable water isotope variation in a Central Andean watershed dominated by glacier- and snowmelt. Hydrology and Earth System Sciences Discussions. 2012;**9**:12227-12269. DOI: 10.5194/hess-17-1035-2013

[6] Peiffer Y, Taran A, Lounejeva E, Solís-Pichardo G, Rouwet D, Bernard-Romero D. Tracing thermal aquifers of El Chichón volcano–hydrothermal system (México) with 87Sr/86Sr, Ca/Sr and REE. Journal of Volcanology and Geothermal Research. 2011;**205**(3–4):55-66. DOI: 10.1016/j.jvolgeores.2011.06.004

[7] Gmati S, Tase N, Tsujimura M, Tosaki Y. Aquifers Interaction in the southwestern Foot of Mt. Fuji, Japan, Examined through Hydrochemistry and Statistical Analyses. Hydrological Research Letters. 2011;**5**:58-63. DOI: 10.3178/hrl.5.58

[8] Andermann C, Longuevergne L, Bonnet S, Crave A, Davy P, Gloaguen R. Impact of transient groundwater storage on the discharge of Himalayan rivers. Nature Geoscience. 2012;**5**(2):127-132. DOI: 10.1038/ngeo1356

[9] Muñoz-Villers L, McDonnell J. Runoff generation in a steep, tropical montane cloud forest catchment on permeable volcanic substrate. Water Resources Research. 2012;**48**:W09528. DOI: 10.1029/2011WR011316

[10] Arumí JL, Oyarzún R, Muñoz E, Rivera D, y Aguirre E. Caracterización de dos grupos de manantiales en el río Diguillín, Chile, Water Science and Technology 2014;**5**(6):151-158

[11] Arumí JL, Maureira H, Souvignet M, Pérez C, Rivera D, Oyarzún R. Where does the water go? Understanding geohydrological behaviour of Andean catchments in south-central Chile. Hydrological Sciences Journal. 2016;**61**(5):844-855. DOI: 10.1080/02626667.2014.934250

[12] Muñoz E, Arumí JL, Wagener T, Oyarzún R, Parra V. Unraveling complex hydrogeological processes in Andean basins in south-central Chile: An integrated assessment to understand hydrological dissimilarity. Hydrological Processes. 2016;**30**(26):4934-4943. DOI: 10.1002/hyp.11032

[13] CNR. Estudio integral de riego Proyecto Itata, Volumen II. Santiago, Chile: Comisión Nacional de Riego; 1994. p. 533

[14] Dixon H, Mick J, Murphy D, Sparks S, Chávez R, Naranjo J, et al. The geology of Nevados de Chillán volcano, Chile. Revista Geologica de Chile. 1999; **26**(2):227-253

[15] Naranjo J, Gilbert J, Sparks R. Geología del complejo volcánico Nevados de Chillán, Región del Biobío (Geology of the volcanic complex Nevados del Chillán). Servicio Nacional de Geología y Minería, Carta Geológica de Chile, Serie Geología Básica 114. 2008; 28. mapa escala 1:50.000

[16] Oyarzún R, Jofré E, Maturana H, Oyarzún J, Aguirre E. Use of $_{222}$Radon as a simple tool for surface water–groundwater connectivity assessment: a case study in the arid Limarí basin, north-central Chile. Water Environment Journal. 2014;**28**(3): 418-422

[17] Rango A, Martinec J. Revisiting the degree-day method for snowmelt computations. Journal of the American Water Resources Association. 1995; **31**(4):657-669

[18] Parisi S, Paternoster M, Kohfahl C, Pekdeger A, Meyer H, Wolfgang H, et al. Groundwater recharge areas of a volcanic aquifer system inferred from hydraulic, hydrogeochemical and stable isotope data: Mount Vulture, southern Italy. Hydrogeology Journal. 2011;**19**: 133-153. DOI: 10.1007/s10040-010-0619-8

[19] Manga M. A model for discharge in spring-dominated streams and implications for the transmissivity and recharge of quaternary volcanics in the Oregon Cascades. Water Resources Research. 1997;**33**(8):1813-1822

[20] Manga M. On the timescales characterizing groundwater discharge at springs. Journal of Hydrology. 1999; **219**(1):56-69. DOI: 10.1016/S0022-1694(99)00044-X

[21] Fry L, David W, Watkins N, Reents M, Rowe D, Mihelcic J. Climate change and development impacts on the sustainability of spring-fed water supply systems in the Alto Beni region of Bolivia. Journal of Hydrology. 2012; **468–469**:120-129. DOI: 10.1016/j.jhydrol.2012.08.023

[22] Müller B, Scheytt T, Asbrand M. Hydrogeology Journal. 2012;**20**:1117. DOI: 10.1007/s10040-012-0852-4

[23] Battaglin W, Bradley P, Iwanowicz L, Journey C, Walsh H, Blazer V. Pharmaceuticals, hormones, pesticides, and other bioactive contaminants in water, sediment, and tissue from Rocky Mountain National Park, 2012–2013. Science of the Total Environment. 2018; **643**:651-673. DOI: 10.1016/j.scitotenv.2018.06.150

[24] Sui Q, Cao X, Lu S, Zhao W, Qiu Z, Yu G. Occurrence, sources and fate of pharmaceuticals and personal care products in the groundwater: A review. Emerging Contaminants. 2015;**1**(1): 14-24. DOI: 10.1016/j.emcon.2015.07.001

Chapter 3

Development of a Method for Prediction of Risk of Surface and Groundwater Contamination with Pesticides and Their Dangerous Aspects for Human Health

Anna Antonenko, Olena Vavrinevych, Maria Korshun and Sergiy Omelchuk

Abstract

The probability of groundwater contamination is high enough because groundwater has different origins: a majority of them are formed by atmospheric precipitation filtration through soil layer or due to condensation of water vapors directly into the ground. Pesticides could be one of such hazardous groundwater pollutants. We developed two methods for the hazardous effect on human organism while consuming contaminated water prediction: risk acceptance assessment and integral groundwater contamination hazard index (IGCHI) evaluation in points according to special scale.

Keywords: groundwater, surface water, hazard, pesticide, leaching, health

1. Introduction

Growing of world population, agriculture, and industrial development led to the increase of ecotoxicants in environmental pollution. Among these ecotoxic substances, pesticides have a special place [1, 2]. Migrating through the soil profile, pesticides create the danger of groundwater contamination that requires their constant control and monitoring [1, 2]. Some older and cheap pesticides, whose application is forbidden in developed countries but are still used in a lot of developing countries, can persist in soil, ground, and surface water for years [3].

At the present time, around 65% of European and 70% of Ukrainian rural and urban population have been using ground (shaft wells) and middle water (artesian wells) for drinking.

As groundwater forms in two ways, (1) water from atmosphere precipitations filtrates through soil or (2) condensation of vapors into the ground, the possibility of groundwater chemical contamination is rather high [4].

That is why prediction of the risk of groundwater contamination with different classes of pesticides, as well as hygienic assessment of their impact on public health is very actual nowadays.

2. Prediction of the risk of ground and surface water contamination with pesticides and its danger to human health in areas with irrigation farming

The prediction of migration opportunities in groundwater of pesticides in different soil and climatic conditions could be carried out by a number of indices.

For example, leaching potential index [groundwater ubiquity score (GUS)] [4] is calculated using the below formula:

$$GUS = \log \tau_{50} \times [4 - \log K_{oc}],$$

where τ_{50}—half-life in soil, days; and

K_{oc}—sorption coefficient of organic carbon.

For the assessment of GUS values, we have used net approach: probability of pesticide leaching into groundwater is present (GUS > 2,8); probability of pesticide leaching into groundwater is possible (GUS < 1,8); pesticides possibly not leached into groundwater (GUS = 1,8–2,8) [5].

US Environmental Protection Agency (EPA) has developed SCI-GROW screening method for the determination of maximum pesticide concentration in groundwater [6], and this model is widely used. SCI-GROW index counts the substance's half-life period in soil, organic carbon sorption coefficient, and pesticide application rate and frequency. The calculation gives the highest possible groundwater concentration of substance in mg/l.

Unfortunately, GUS index has disadvantages. For example, not all significant parameters that can influence the behavior of pesticide in the system "groundwater" are taking into account; run-off to surface water cannot be assessed using this value.

LEACH index is better. It determines also the possibility of river contamination and takes into account the maximum number of parameters that can influence the transition of pesticides from soil into other mediums.

The index of potential contamination of groundwater and river water LEACH was calculated according to the below formula [7]:

$$LEACH_{\text{mod.}} = \frac{S_w \times DT_{50field}}{K_{oc}},$$

where S_w—water solubility, mg/l;

$DT_{50\ field}$—half-life period substances in the soil in natural conditions, day; and

K_{oc}—organic carbon (o.c.) sorption coefficient, ml/g o.c.

Evaluation of the index: 0,0–1,0-low risk of pollution (3 class), 1,1–2,0-average (moderate) risk (2 class), and >2,0-high risk (1 class).

But all the above listed indices characterize only the potential of pesticide penetration into groundwater and surface water without the possibility of evaluation of risk for human organism while consumption of contaminated water.

So, method of comprehensive assessment of pesticides leaching into the water possible adverse effects on humans developed by us has been used for the SCI-GROW evaluation [8]. The principle of complex hygienic regulation takes into account the possibility of pesticide intake through inhalation, with drinking water and food and its safe levels, is in the base of this method. Pesticide acceptable daily intake with water (PADIW) compares with pesticide maximum possible daily intake with water (PMDIW), which ways of calculations in 3 steps is given below (**Figure 1**).

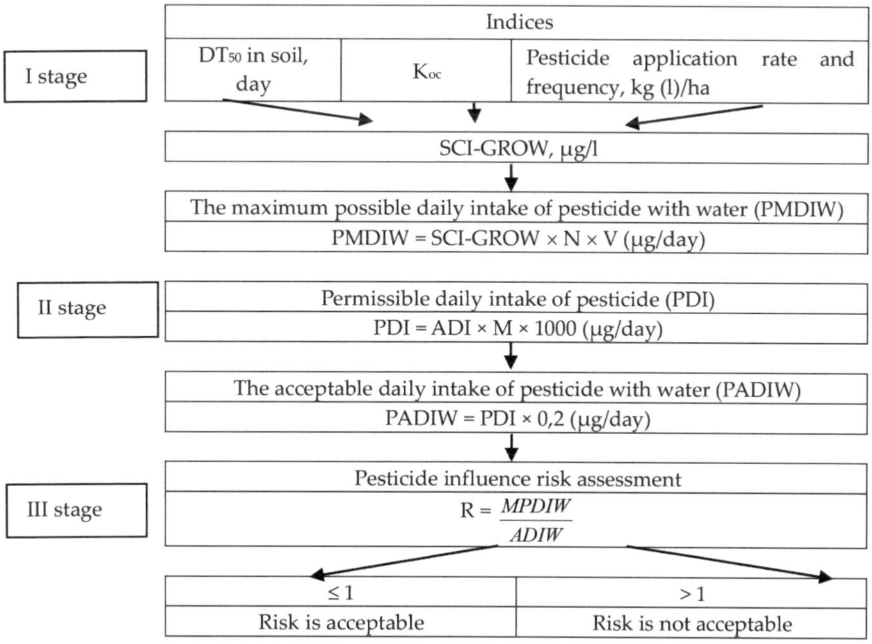

Figure 1.
A method for assessing the risk of adverse effects of pesticides on human health when consuming contaminated water. Notes: SCI-GROW—screening concentrations of pesticides in groundwater, μg/l; V—daily intake of water by human, l (3 l—in temperate climate, 5–10 l—in hot climate); ADI—acceptable daily intake of pesticide, mg/kg; M—average weight of person (60 kg); 1000—factor for conversion in micrograms.

Initially, one needs to calculate the SCI-GROW using computer program from EPA official Website. This indicator is based on the actual results of field studies; therefore, it gives the most realistic values. In order to obtain the maximum possible value of pesticide intake with water (PMDIW) by humans, SCI-GROW index is multiplied by the average daily consumption of water (for persons living in temperate climate-3 L, for those living in hot climate-5 to 10 L).

To evaluate the obtained indicator, it is necessary to calculate the permissible level of pesticide intake with water (PADIW). For this, firstly, the allowable daily dose (ADI) must be multiplied by the average weight of a person (M) (60 kg for nonprofessional contingents and 70 kg for professionals). Based on the principles of complex hygienic regulation, the amount of pesticide that entered the human body with water should not exceed 20% of the permissible daily intake. Therefore, the indicator obtained earlier is multiplied by 0.2.

Finally, the values of PMDIW and PADIW should be compared (R). If the R value is ≤1, risk is considered to be acceptable; and if R > 1, risk is not acceptable.

Also, we recommend integrated assessment of the potential hazard of pesticide exposure on the human organism when consuming contaminated drinking water to use the scale with four gradations (**Figure 2**). The scale includes three indices: LEACH, τ_{50} in water, and acceptable daily intake (ADI) [9, 10].

These three indicators mostly reflect the danger of a pesticide, when ingested with water. LEACH displays the maximum possible risk of contamination of water supply sources, both underground and surface, taking into account, the physical properties of the main pesticide and stability in soil. τ_{50} displays the possibility and duration of the presence of the pesticide in the potentially drinking water. ADI, the main and integral pesticide toxicity index, shows the possibility of the realization of the toxic effects of a substance, when it is present in water for a long period.

Indices	$LEACH_{mod.} = \frac{S_w \times \tau_{50field}}{K_{oc}}$, where Sw – water solubility, mg/l; τ_{50} field – half-life period substances in the soil in natural conditions, day; Koc – organic carbon (o.c.) sorption coefficient, ml/g o.c.	τ_{50} in water, day	Acceptable daily intake (ADI), mg/kg

Points, depends on index value			
Points	LEACH	τ_{50} in water	ADI
1	<0,01	<5	>0,02
2	0,01-0,1	5-10	0,0051–0,02
3	0,11-1,0	11-30	0,0021–0,005
4	>1,0	>30	≤0,002

Integral groundwater contamination hazard index (IGCHI)

IGCHI = LEACH + τ_{50} + ADI (points)

IGCHI evaluation and determination of hazard class	
3 and 4 points	substances are low hazardous for human (4 class)
5 and 6 points	moderately hazardous (3 class)
7 and 8 points	hazardous (2 class)
9 and 10 points	highly hazardous (1B class)
11 and 12 points	extremely hazardous (1A class)

Figure 2.
Method of hazard prediction of contaminated water by pesticide water effect on human body. Note. Evaluation of the LEACH index: 0,0–1,0—low risk of pollution (3 class), 1,1–2,0—average (moderate) risk (2 class), and >2,0—high risk (1 class).

For testing proposed by us, methods of risk assessment of pesticide-contaminated drinking water, we have studied widely used in agriculture representatives of the most perspective chemical classes of herbicides, fungicides, and insecticides (**Tables 1–3**). The main physical and chemical properties of studied compounds are given in **Table 1–3**.

The conditions of studied pesticides application and stability are given in **Table 4**.

International IUPAC classification [15] was used to assess the literature data about the stability and mobility of substances in the soil. The first includes three classes: 1-highly persistent (with DT_{50} more than 100 days), 2-moderately persistent (30–100 days), and 3-low persistent (less than 30 days).

According to IUPAC classification [15], most of fungicides and insecticides by persistence in soil may be attributed to moderately persistent (2 class); all herbicides, to low persistent (3 class). Exceptions are highly persistent insecticides, imidacloprid and chlorantraniliprole; fungicides, sedaxane, boscalid, fluxapyroxad, and azoxystrobin; and moderately persistent herbicides, triasulfurone and imazethapyr (**Table 3**). It should be noted that these literature data are very average. For example, in the soil and climatic conditions of the southern and southeastern European countries, including Ukraine, the transformation of the studied substances occurs much faster due to microbiological degradation (typical for these regions, black soils are rich in microflora) [8].

Trade name	Chemical name (IUPAC)	lg K_{ow}	Solubility in water, mg/l	K_{oc}
Triazoles				
Difenoconazole	3-chloro-4-[(2RS.4RS;2RS.4SR)-4-methyl-2-(1H-1.2.4-triazol-1-ylmethyl)-1.3-dioxolan-2-yl]phenyl 4-chlorophenyl ether	4.2	15.0	3760
Tebuconazole	(RS)-1-p-chlorophenyl-4.4-dimethyl-3-(1H-1.2.4-triazol-1-ylmethyl)pentan-3-ol	3.7	32.0	769
Penconazole	(RS)-1-[2-(2.4-dichlorophenyl)pentyl]-1H-1.2.4-triazole	3.72	73.0	2205
Strobilurines				
Pyraclostrobin	methyl {2-[1-(4-chlorophenyl)pyrazol-3-yloxymethyl]phenyl}(methoxy)carbamate	3.99	1.9	9304
Azoxystrobin	methyl (E)-2-{2-[6-(2-cyanophenoxy)pyrimidin-4-yloxy]phenyl}-3-methoxyacrylate	2.5	6.7	589
Trifloxystrobin	methyl (E)-methoxyimino-{(E)-α-[1-(α.α.α-trifluoro-m-tolyl)ethylideneaminooxy]-o-tolyl}acetate	4.5	0.61	2377
Ethylene-bis-dithiocarbamate				
Metiram	zinc ammoniate ethylenebis(dithiocarbamate) - poly (ethylenethiuram disulfide)	1.76	2.0	998
Mancozeb	manganese ethylenebis(dithiocarbamate) (polymeric) complex with zinc salt	1.33	6.2	500,000
Cyanopyrrole				
Fludioxonil	4-(2.2-difluoro-1.3-benzodioxol-4-yl)-1H-pyrrole-3-carbonitrile	4.12	1.8	145,600
Anilidepyrimidines				
Cyprodinil	4-cyclopropyl-6-methyl-N-phenylpyrimidin-2-amine	4.5	13.0	2277
Pyrimethanil	N-(4.6-dimethylpyrimidin-2-yl)aniline	2.84	0.121	301
Valifenale	methyl N-(isopropoxycarbonyl)-L-valyl-(3RS)-3-(4-chlorophenyl)-β-alaninate	3.11	24.1	1686
Pyrazolecarboxamides				
Fluxapyroxad	3-(difluoromethyl)-1-methyl-N-(3′.4′.5′-trifluorobiphenyl-2-yl)pyrazole-4-carboxamide	3.13	3.44	728
Isopyrazam	mixture of 2 isomers 3-(difluoromethyl)-1-methyl-N-[(1RS.4SR.9RS)-1.2.3.4-tetrahydro-9-isopropyl-1.4-methanonaphthalen-5-yl]pyrazole-4-carboxamide and 2 isomers 3-(difluoromethyl)-1-methyl-N-[(1RS.4SR.9SR)-1.2.3.4-tetrahydro-9-isopropyl-1.4-methanonaphthalen-5-yl]pyrazole-4-carboxamide	4.25	0.55	2416
Penthiopyrad	(RS)-N-[2-(1.3-dimethylbutyl)-3-thienyl]-1-methyl-3-(trifluoromethyl)pyrazole-4-carboxamide	4.62	1.375	804
Sedaxane	mix of: trans-isomers 2′-[(1RS.2SR)-1.1′-bicycloprop-2-yl]-3-(difluoromethyl)-1-methyl-1H-pyrazole-4-carboxanilide and 2 cis-isomers 2′-[(1RS.2RS)-1.1′-bicycloprop-2-yl]-3-(difluoromethyl)-1-methyl-1H-pyrazole-4-carboxanilide	3.3	14.0	534

Trade name	Chemical name (IUPAC)	lg K_{ow}	Solubility in water, mg/l	K_{oc}
Anilides				
Benalaxyl-M	methyl N-(phenylacetyl)-N-(2.6-xylyl)-D-alaninate	3.67	33.0	7175
Boscalid	2-chloro-N-(4′-chlorobiphenyl-2-yl)nicotinamide	2.96	4.6	772

Table 1.
Physical and chemical properties of the studied fungicides [11, 12].

Trade name	Chemical name (IUPAC)	lg K_{ow}	Solubility in water, mg/l	K_{oc}
Organophosphates				
Chlorpyrifos	O.O-diethyl O-3.5.6-trichloro-2-pyridyl phosphorothioate	4.7	1.05	8151
Dimethoate	2-dimethoxyphosphinothioylthio-N-methylacetamide	0.704	39,800	28.3
Pyrethroid				
Bifenthrin	2-methyl-3-phenylbenzyl (1RS)-cis-3-(2-chloro-3.3.3-trifluoroprop-1-enyl)-2.2-dimethylcyclopropanecarboxylate	6.6	0.001	236,610
Cyperme-thrin	(RS)-α-cyano-3-phenoxybenzyl (1RS.3RS;1RS.3SR)-3-(2.2-dichlorovinyl)-2.2-dimethylcyclopropanecarboxylate	5.3	0.009	156,250
Alpha-cyperme-thrin	Racemate comprising (R)-α-cyano-3-phenoxybenzyl (1S)-cis-3-(2.2-dichlorovinyl)-2.2-dimethylcyclopropanecarboxylate and (S)-α-cyano-3-phenoxybenzyl (1R)-cis-3-(2.2-dichlorovinyl)-2.2-dimethylcyclopropanecarboxylate	6.94	0.004	57,889
Lambda-cyhalothrin	(R)-a-cyano-3-phenoxybenzyl (1S)-cis-3-[(Z)-2-chloro-3.3.3-trifluoropropenyl]-2.2-dimethylcyclopropanecarboxylate and (S)-a-cyano-3-phenoxybenzyl (1R)-cis-3-[(Z)-2-chloro-3.3.3-trifluoropropenyl]-2.2-dimethylcyclopropanecarboxylate	5.5	0.005	283,707
Neonicotinoid				
Thiame-thoxam	(EZ)-3-(2-chloro-1.3-thiazol-5-ylmethyl)-5-methyl-1.3.5-oxadiazinan-4-ylidene(nitro)amine	−0.13	4100	56.2
Imidaclo-prid	(E)-1-(6-chloro-3-pyridylmethyl)-N-nitroimidazolidin-2-ylideneamine	0.57	610	225
Pyrazolium				
Tebufen-pyrad	N-(4-tert-butylbenzyl)-4-chloro-3-ethyl-1-methylpyrazole-5-carboxamide	4.93	2.39	5992
Chlorantra-niliprole	3-bromo-4′-chloro-1-(3-chloro-2-pyridyl)-2′-methyl-6′-(methylcarbamoyl)pyrazole-5-carboxanilide	4.22	0.88	362
Benzoylurea				
Novaluron	(RS)-1-[3-chloro-4-(1.1.2-trifluoro-2-trifluoromethoxyethoxy)phenyl]-3-(2.6-difluorobenzoyl)urea	4.3	0.003	9598

Table 2.
Physical and chemical properties of the studied insecticides [11].

Trade name	Chemical name (IUPAC)	lg K_{ow}	Solubility in water, mg/l	K_{oc}
Chloroacetamides				
Acetochlore	2-chloro-N-ethoxymethyl-6′-ethylacet-o-toluidide	4.14	282	156
Dimetachlor	2-chloro-N-(2-methoxyethyl)acet-2′.6′-xylidide	2.17	2300	69
Propizochlor	2-chloro-6′-ethyl-N-isopropoxymethylacet-ortho-toluidide	3.3	90.8	291
S-metolachlor	Mix of: (aRS.1S)-2-chloro-6′-ethyl-N-(2-methoxy-1-methylethyl)acet-o-toluidide and (aRS.1R)-2-chloro-6′-ethyl-N-(2-methoxy-1-methylethyl)acet-o-toluidide	3.05	480	226.1
Metasachlor	2-chloro-N-(pyrazol-1-ylmethyl)acet-2′.6′-xylidide	2.49	450	54
Sulfonil-carbonyl-triazolinone				
Thiencarbazon-methyl	Methyl 4-[(4.5-dihydro-3-methoxy-4-methyl-5-oxo-1H-1.2.4-triazol-1-yl)carbonylsulfamoyl]-5-methylthiophene-3-carboxylate	−1.98	436	100
Oxazoles				
Topramezone	[3-(4.5-dihydro-1.2-oxazol-3-yl)-4-mesyl-o-tolyl](5-hydroxy-1-methylpyrazol-4-yl)methanone	−1.52	100,000	15.0–296.7
Isoxaflutole	(5-cyclopropyl-1.2-oxazol-4-yl)(α.α.α-trifluoro-2-mesyl-p-tolyl)methanone	2.32	6.2	112
Triketones				
Mesotrione	2-(4-mesyl-2-nitrobenzoyl)cyclohexane-1.3-dione	0.11	160	80
Sulfonylurea				
Foramsulfu-rone	1-(4.6-dimethoxypyrimidin-2-yl)-3-[2-(dimethylcarbamoyl)-5-formamidophenylsulfonyl] urea	−0.78	3293	78
Iodsulfurone methyl-sodium	Sodium ({[5-iodo-2-(methoxycarbonyl)phenyl] sulfonyl} carbamoyl) (4-methoxy-6-methyl-1.3.5-triazin-2-yl)azanide	1.59	25,000	45
Phosphonoglycine				
Glyphosate	N-(phosphonomethyl)glycine	−3.2	10,500	21,699
Sulfonylurea with triazine heterrocycle				
Tritosulfuron	N-{[4-methoxy-6-(trifluoromethyl)-1.3.5-triazin-2-yl]carbamoyl}-2-(trifluoromethyl)benzene-1-sulfonamide	2.93	78.3	7.5
Prosulfuron	1-(4-methoxy-6-methyl-1.3.5-triazin-2-yl)-3-[2-(3.3.3-trifluoropropyl)phenylsulfonyl]urea	1.5	4000	14.2
Metsulfuron-methyl	Methyl 2-(4-methoxy-6-methyl-1.3.5-triazin-2-ylcarbamoylsulfamoyl)benzoate	−1.87	2790	12.0
Triasulfuron	1-[2-(2-chloroethoxy)phenylsulfonyl]-3-(4-methoxy-6-methyl-1.3.5-triazin-2-yl)urea	−0.59	815	60
Tribenuron-methyl	Methyl 2-[4-methoxy-6-methyl-1.3.5-triazin-2-yl (methyl)carbamoylsulfamoyl]benzoate	0.38	2483	35
Sulfonylurea with pyrimidine heterocycle				
Rimsulfuron	1-(4.6-dimethoxypyrimidin-2-yl)-3-(3-ethylsulfonyl-2-pyridylsulfonyl)urea	−1.46	7300	50.3

Trade name	Chemical name (IUPAC)	lg K_{ow}	Solubility in water, mg/l	K_{oc}
Nicosulfuron	2-[(4.6-dimethoxypyrimidin-2-ylcarbamoyl)sulfamoyl]-N.N-dimethylnicotinamide	0.61	7500	30
Chlorimuron-ethyl	Ethyl 2-(4-chloro-6-methoxypyrimidin-2-ylcarbamoylsulfamoyl)benzoate	0.11	1200	106
Imidazolinone				
Imazapyr	2-[(RS)-4-isopropyl-4-methyl-5-oxo-2-imidazolin-2-yl]nicotinic acid	0.11	9740	125
Imazamox	2-[(RS)-4-isopropyl-4-methyl-5-oxo-2-imidazolin-2-yl]-5-methoxymethylnicotinic acid	5.36	626,000	11.6
Imazethapyr	5-ethyl-2-[(RS)-4-isopropyl-4-methyl-5-oxo-2-imidazolin-2-yl]nicotinic acid	1.49	1400	52
Pyrimidinyl carboxy compound				
Bispyribac-sodium	Sodium 2.6-bis(4.6-dimethoxypyrimidin-2-yloxy) benzoate	−1.03	64,000	302
Semicarbazone				
Diflufenzopyr	2-{(EZ)-1-[4-(3.5-difluorophenyl)semicarbazono] ethyl}nicotinic acid	1.09	5850	87

Table 3.
Physical and chemical properties of the studied herbicides [11].

Half of the studied herbicides and insecticides are resistant or highly resistant in water, as they are poorly decomposed by photolysis and hydrolysis. Fungicides are much less resistant (**Table 3**).

It was found that according to GUS index, there is no risk of leaching into groundwater for most of the studied herbicides; for the rest, it is low. Only for one fungicide (topramezone) and most of insecticides, the risk of groundwater leaching is high (**Table 5**). It could be explained by their high toxicity (very low ADI values) and relatively high persistency in soil and water (**Table 4**).

The calculated maximum possible concentrations of the studied fungicides, herbicides, and insecticides SCI-GROW in groundwater indicate that the risk to humans when consuming such water is acceptable (**Table 5**). SCI-GROW values exceed 1 μg/l only for triasulfurone, imazamox, imazethapyr, and chlorantraniliprole. But the high risk will not be realized as shown in **Table 5**; IGHI values for these pesticides are 7, 6, 6, and 7, respectively.

According to IGCHI index, fungicides, penconazole and azoxystrobin; herbicides, dimetachlor, propizochlor, s-metolachlor, foramsulfurone, glyphosate, and rimsulfuron are less hazardous for human organism in case of consuming contaminated water. Fungicides, difenoconazole, pyraclostrobin, trifloxystrobin, metiram, mancozeb, fludioxonil, valifenale, fluxapyroxad, isopyrazam, penthiopyrad, and boscalid; herbicides, metazachlor, thiencarbazone-methyl, isoxaflutole, iodsulfuron methyl-sodium, metsulfuron-methyl, nicosulfuron, chlorimuron-ethyl, imazapyr, imazamox, imazethapyr, and diflufenzopyr; insecticides, thiamethoxam and imidacloprid are moderately hazardous (**Table 5**). Only insecticides, chlorpyrifos, bifenthrin, lambda-cyhalothrin, and tebufenpyrad are highly and extremely hazardous because of their high toxicity and water pollution possibility. Rest of the studied compounds is hazardous (2 class) to human organism.

Active ingredient (a.i.)	Maximum application rate of a.i., kg/ha	DT_{50} soil, day	DT_{50} water, day	Acute oral LD_{50} (mg/kg) (rat)	ADI, mg/kg	PDI, mg/day
Fungicides						
Difenoconazole	0.250	85 (20–265)	3.0	1453	0.01	0.6
Tebconazole	0.175	47.1 (25.8–91.6)	42.6	1700	0.03	1.8
Penconazole	0.160	90 (22–115)	2.0	>2000	0.03	1.8
Pyraclostrobin	0.100	32 (8–55)	2.0	>5000	0.03	1.8
Azoxystrobin	0.200	180.7 (120.9–261.9)	6.1	>5000	0.20	12.0
Trifloxystrobin	0.175	7 (2–12)	1.1	>5000	0.10	6.0
Metiram	1.750	7 (~7)	0.7	>5000	0.03	1.8
Mancozeb	1.625	18 (1)	0.2	>5000	0.05	3.0
Fludioxonil	0.250	20.5 (8–43)	2.0	>5000	0.37	22.2
Cyprodinil	0.375	45 (11–98)	12.5	>2000	0.03	1.8
Pyrimethanil	0.480	29.5 (23–54)	16.5	4150	0.17	10.2
Valifenale	0.306	1.9–12.0 hours	5.0	>5000	0.07	4.2
Fluxapyroxad	0.126	151 (53–424)	4.4	>2000	0.02	1.2
Isopyrazam	0.450	72 (9.11–173)	2.3	2000	0.03	0.6
Penthiopyrad	0.390	47 (0.8–33.3)	9.9	>2000	0.10	6.0
Sedaxane	0.025	170 (54.6–188.0)	17.3	>2000	0.10	6.0
Benalaxyl-M	0.400	44 (36–124)	38.0	>2000	0.01	2.1
Boscalid	0.668	118 (28–208)	9.0	>5000	0.04	2.4
Herbicides						
Acetochlore	2.700	12.1 (7.0–17.0)	40.5	1929	0.0036	0.220
Dimetachlor	1.200	3.2 (2.3–15.6)	10.0	1600	0.1	6.000
Propizochlor	2.160	7.63 (10.0–15.0)	8.5	2290	0.025	1.500
S-metolachlor	1.920	21.0 (11.0–31.0)	9.0	2577	0.1	6.000
Metasachlor	1.250	6.8 (26.0–114.0)	216.0	3480	0.08	4.800
Thiencarbazone-methyl	0.045	17.0 (14.0–45.0)	118	>2000	0.23	13.80
Topramezone	0.075	26.1 (10.8–69.3)	30	>2000	0.001	0.060
Isoxaflutole	0.1125	1.3 (0.5–2.4)	11	>5000	0.02	1.200
Mesotrione	0.110	5.0 (3.0–7.0)	>30	>5000	0.01	0.600
Foramsulfurone	0.045	5.5 (12.0–15.0)	10	>5000	0.25	30.00
Iodsulfurone methyl-sodium	0.0015	3.2 (0.8–10.3)	31	2448	0.03	1.800
Glyphosate	1.6654	23.79 (5.7–40.9)	2.5	>2000	0.3	18.00
Tritosulfuron	0.0500	12 (3–21)	20.0	>5000	0.15	9.0
Prosulfuron	0.0150	11.9 (3.8–38.9)	173.0	546	0.02	1.2
Metsulfuron-methyl	0.0060	13.3 (7.3–37.1)	224.3	>5000	0.22	13.2
Triasulfuron	0.0062	38.5 (16.1–92.4)	217.0	>5000	0.01	0.6

Active ingredient (a.i.)	Maximum application rate of a.i., kg/ha	DT_{50} soil, day	DT_{50} water, day	Acute oral LD_{50} (mg/kg) (rat)	ADI, mg/kg	PDI, mg/day
Tribenuron-methyl	0.0188	10 (5–20)	139.0	>5000	0.01	0.6
Rimsulfuron	0.0125	10.8 (5.6–17.7)	6.0	>5000	0.1	6.0
Nicosulfuron	0.0600	19.3 (8.9–63.3)	65.0	>5000	2.0	120.0
Chlorimuron-ethyl	0.0094	28 (14–42)	21.0	>4102	0.02	1.2
Imazapyr	0.0550	11 (5.9–16.5)	30.0	>2000	2.5**	156.0
Imazamox	0.0400	16.7 (8.1–14.0)	233	>5000	9.0	540.0
Imazethapyr	0.1200	51.0 (14.0–290.0)	520	>5000	0.44	26.4
Bispyribac-sodium	0.0450	6.3 (2.1–7.6)	35.3	2635	0.01	0.6
Diflufenzopyr	0.0680	4.5 (8.0–18.0)	24.0	>5000	0.26	15.6
Insecticides						
Chlorpyrifos	0.720	27.6 (0.32–88.9)	36.5	66	0.001	0.060
Dimethoate	0.600	7.2 (4.6–9.8)	15.5	245	0.001	0.060
Bifenthrin	0.060	86.8 (5.4–267.0)	161.0	54.5	0.015	0.900
Cypermethrin	0.075	21.9 (14.0–199.0)	17.0	287	0.05	3.000
Alpha-cypermethrin	0.030	42.6 (14.0–112.0)	21.0	40	0.015	0.090
Lambda-cyhalothrin	0.0424	26.9 (10.1–47.5)	15.1	56	0.0025	0.150
Thiamethoxam	0.150	39.0 (7.0–72.0)	40.0	>1563	0.026	1.560
Imidacloprid	0.060	174 (104.0–228.0)	129.0	131	0.06	3.600
Tebufenpyrad	0.160	4.5 (0.05–22.4)	90.0	>202	0.01	0.600
Chlorantraniliprole	0.050	204.0 (123.0–561.0)	170.0	>5000	1.56	93.60
Novaluron	0.060	96.5 (33.0–160.0)	17.5	>5000	0.01	0.600

Note. PDI: permissible daily intake of pesticide.
***The table gives the initial data for the evaluation and shows the results of calculations of the index proposed by us (testing the method).*

Table 4.
The conditions of studied pesticides' application and stability [9–11, 13, 14].

Active ingredient	GUS	SCI-GROW (μg/l)	Leach		IGCHI	
			Value	Class	Value	Class
Fungicides						
Difenoconazole	0.9	1.79×10^{-2}	3.391×10^{-1}	3	6	3
Tebconazole	2.0	2.77×10^{-1}	$1.9599 \times 10^{+0}$	2	7	2
Penconazole	1.36	3.38×10^{-2}	$2.9796 \times 10^{+0}$	1	3	4
Pyraclostrobin	0.05	5.52×10^{-3}	6.500×10^{-3}	3	5	3
Azoxystrobin	2.60	1.98×10^{-1}	$2.0555 \times 10^{+0}$	1	4	4
Trifloxystrobin	0.53	1.43×10^{-5}	1.800×10^{-3}	3	5	3
Metiram	0.00	5.35×10^{-3}	1.40×10^{-2}	3	5	3

Active ingredient	GUS	SCI-GROW (μg/l)	Leach		IGCHI	
			Value	Class	Value	Class
Mancozeb	−1.00	2.84×10^{-6}	2.000×10^{-4}	3	5	3
Fludioxonil	−2.48	5.35×10^{-3}	3.000×10^{-4}	3	5	3
Cyprodinil	1.01	2.33×10^{-2}	2.569×10^{-1}	3	7	2
Pyrimethanil	2.65	1.90×10^{-1}	1.19×10^{-2}	3	7	2
Valifenale	−0.68	1.97×10^{-5}	0.0071×10^{-3}	3	6	3
Fluxapyroxad	2.57	1.85×10^{-1}	7.135×10^{-1}	3	6	3
Isopyrazam	1.47	4.01×10^{-2}	1.64×10^{-2}	3	5	3
Penthiopyrad	2.33	1.31×10^{-1}	1.57×10^{-2}	3	6	3
Sedaxane	2.59	1.85×10^{-4}	$4.46 \times 10^{+0}$	1	8	2
Benalaxyl-M	0.41	9.34×10^{-3}	2.024×10^{-1}	3	8	2
Boscalid	2.56	2.10×10^{-1}	7.031×10^{-1}	3	6	3
Herbicides						
Acetochlore	1.58	2.58×10^{-2}	$3.073 \times 10^{+1}$	1	8	2
Dimetachlor	1.76	8.68×10^{-3}	$5.20 \times 10^{+2}$	1	4	4
Propizochlor	1.36	1.26×10^{-2}	$4.68 \times 10^{+0}$	1	4	4
S-metolachlor	1.91	4.85×10^{-2}	$6.581 \times 10^{+1}$	1	4	4
Metasachlor	2.17	4.73×10^{-2}	$9.50 \times 10^{+2}$	1	6	3
Thiencarbazone-methyl	2.46	1.03×10^{-1}	$1.962 \times 10^{+2}$	1	6	3
Topramezone	5.06	0.567×10^{-1}	$2.336 \times 10^{+4}$	1	8	2
Isoxaflutole	0.59	1.28×10^{-3}	$9.244 \times 10^{+2}$	1	6	3
Mesotrione	1.47	4.13×10^{-3}	$1.400 \times 10^{+1}$	1	7	2
Foramsulfurone	1.56	4.63×10^{-3}	$6.333 \times 10^{+2}$	1	4	4
Iodsulfurone methyl-sodium	0.71	1.64×10^{-3}	$5722 \times 10^{+3}$	1	6	3
Glyphosate	−0.36	5.35×10^{-3}	$1.979 \times 10^{+1}$	1	3	4
Tritosulfuron	2.81	2.43×10^{-1}	4.00×10^{-2}	3	7	2
Prosulfuron	5.11	$4.17 \times 10^{+0}$	$3.61 \times 10^{+0}$	1	7	2
Metsulfuron-methyl	3.99	6.89×10^{-1}	$8.626 \times 10^{+3}$	1	6	3
Triasulfuron	5.12	$4.13 \times 10^{+0}$	$1.255 \times 10^{+3}$	1	7	2
Tribenuron-methyl	2.40	4.17×10^{-2}	$1.419 \times 10^{+3}$	1	7	2
Rimsulfuron	3.23	3.17×10^{-1}	$2.569 \times 10^{+3}$	1	4	4
Nicosulfuron	3.25	2.38×10^{-1}	$1.583 \times 10^{+4}$	1	6	3
Chlorimuron-ethyl	3.16	3.55×10^{-1}	$4.755 \times 10^{+2}$	1	6	3
Imazapyr	1.98	4.02×10^{-2}	$1.286 \times 10^{+3}$	1	5	3
Imazamox	6.76	$3.92 \times 10^{+1}$	$2.026 \times 10^{+2}$	1	6	3
Imazethapyr	6.19	$2.59 \times 10^{+1}$	$7.808 \times 10^{+3}$	1	6	3
Bispyribac-sodium	1.68	3.41×10^{-2}	$1.611 \times 10^{+3}$	1	7	2
Diflufenzopyr	2.36	7.85×10^{-2}	$1.210 \times 10^{+3}$	1	5	3
Insecticides						
Chlorpyrifos	0.17	6.45×10^{-3}	1.15×10^{-2}	3	11	1A

Active ingredient	GUS	SCI-GROW (μg/l)	Leach		IGCHI	
			Value	Class	Value	Class
Dimethoate	1.06	2.36×10^{-3}	$1.38 \times 10^{+4}$	1	8	2
Bifenthrin	−2.76	$5.35 \times 10 - ^{3}$	1.13×10^{-6}	3	9	1B
Cypermethrin	−2.19	5.35×10^{-3}	1.15×10^{-5}	3	7	2
Alpha-cypermethrin	−1.53	$5.35 \times 10 - ^{3}$	7.74×10^{-6}	3	8	2
Lambda-cyhalothrin	−3.28	$5.35 \times 10 - ^{3}$	8.37×10^{-7}	3	9	1B
Thiamethoxam	4.69	$3.14 \times 10^{+0}$	$5.25 \times 10^{+3}$	1	6	3
Imidacloprid	3.74	$9.29 \times 10 - ^{1}$	$6.18 \times 10^{+2}$	1	6	3
Tebufenpyrad	0.58	1.11×10^{-2}	8.93×10^{-3}	3	9	1B
Chlorantraniliprole	4.22	$1.86 \times 10^{+0}$	$1.36 \times 10^{+0}$	2	7	2
Novaluron	0.02	5.20×10^{-3}	5.00×10^{-5}	3	8	2

Table 5.
Ground and surface water migration parameters of studied pesticides [8–10, 13, 14].

The estimate presented is approximate. In each particular case, it is necessary to assess the risk of a pesticide when it enters the human body with water separately, taking into account the soil and climatic conditions of the application area, the norms of application, the groundwater depth, and other background factors.

3. Conclusions

1. It was determined that according to IUPAC classification, most of the pesticides pertain to low or moderate in soil, but for some of them, there is a risk of groundwater contamination.

2. Two methods for hazardous effect on human organism while consuming contaminated water prediction were developed by us. For integrated assessment of the potential hazard of pesticide exposure on the human organism when it enters ground and surface waters, we developed integral groundwater contamination hazard index (IGCHI), which includes assessment of three indices: LEACH, τ_{50} in water, and allowable daily intake (ADI) on a scale, which provides four gradations. For the evaluation of the parameters of SCI-GRW, a method of comprehensive assessment including establishment of the maximum possible daily intake of pesticide with water (PMDIW) and subsequently compared with acceptable daily intake of pesticide with water (PADIW) developed by us was used.

3. It was shown that when the human body reaches the majority of investigated compounds, when evaluated using first method, the risk is acceptable. According to the second method, only insecticides were highly or extremely dangerous for the human body while drinking contaminated water. The rest of the compounds are low or moderately hazardous.

Author details

Anna Antonenko[1]*, Olena Vavrinevych[1], Maria Korshun[2] and Sergiy Omelchuk[3]

1 Hygiene and Ecology Department № 1 of O.O. Bogomolets National Medical University, Kyiv, Ukraine

2 Hygiene and Ecology Department № 3 of O.O. Bogomolets National Medical University, Kyiv, Ukraine

3 Hygiene and Ecology Institute of O.O. Bogomolets National Medical University, Kyiv, Ukraine

*Address all correspondence to: antonenko1985@ukr.net

References

[1] Kookana RS, Baskaran S, Naidu R. Pesticide fate and behaviour in Australian soils in relation to contamination and management of soil and water: A review. Australian Journal of Soil Research. 1998;**36**(5):715-764. DOI: 10.1071/S97109

[2] Bartha R, Lanzilotta P, Pramer D. Stability and effect of some pesticides in soil. Applied Microbiology. 1967;**1**:67-75

[3] Pesticides and Water Pollution. Safe Drinking Water Foundation [Internet]. 2018. Available from: https://www.safewater.org/fact-sheets-1/2017/1/23/pesticides [Accessed: October 18, 2018]

[4] Zasypka LG. Health condition of the population in areas of intensive agricultural production. Medical Perspectives. 2011;**XVI**(1):91-96

[5] Gustafson DI. Groundwater ubiquity score: A simple method for assessing pesticide leachability. Environmental Toxicology and Chemistry. 1989;**8**: 339-357. DOI: 10.1002/etc.5620080411

[6] Cohen S. Recent examples of pesticide assessment and regulation under FQPA. Groundwater Monitoring & Remediation. 2000:41-43

[7] Spadotto CA. Screening method for assessing pesticide leaching potential. Pesticidas: Revista de Ecotoxicologia e Meio Ambiente. 2002;**12**:69-78. DOI: 10.5380/pes.v12i0.3151

[8] Vavrinevych OP, Antonenko AM, Omelchuk ST, Korshun MM, Bardov VG. Prediction of soil and ground water contamination with fungicides of different classes according to soil and climate conditions in Ukrain and other European countries. Georgian Medical News. 2015;**5**(242):73-84. DOI: 10.21303/2585-663.2017.00441

[9] Antonenko AM, Vavrinevych OP, Omelchuk ST, Korshun MM. Comparative hygienic evaluation and prediction of hazard to human health of groundwater contamination by herbicides of the most common chemical classes. The unity of science: International Scientific Periodical Journal. 2015:153-157

[10] Antonenko AM, Vavrinevych OP, Omelchuk ST, Korshun MM. Prediction of pesticide risks to human health by drinking water extracted from underground sources. Georgian Medical News. 2015;**7-8**(244–245):99-106

[11] PPDB: Pesticide Properties Data Base [Internet]. 2018. Available from: http://sitem.herts.ac.uk/aeru/footprint/en/ [Accessed: August 25, 2018]

[12] Carisse O, editor. Fungicides. London: InTech; 2010. p. 538

[13] Antonenko AM, Vavrinevych OP, Omelchuk ST, Korshun MM. Comparative hygienic risk assessment of groundwater contamination by herbicides of different chemical classes and hazard prediction for human after consumption of contaminated water. Journal of Education, Health and Sport. 2016;**9**:873-882. DOI: 10.5281/zenodo.161844

[14] Korshun M, Dema O, Kucherenko O, Korshun O, Garkavyi S, Pelio I, et al. Predicting of risks of groundwater and surface water pollution with different classes of herbicides in soil in Eastern Europe climate conditions. Georgian Medical News. 2016;**11**(260):86-90

[15] Categories of Fungicide Solubility, Persistence and Mobility in Soils (adapted from Karmin). US: IUPAC. Pesticide Properties Database; 1997

Section 2

Groundwater Management

Chapter 4

Sustainability of Human, Plant, and Aquatic Life: A Theoretical Discussion from Recharge to Discharge

Joe Magner and Modreck Gomo

Abstract

Groundwater comprises about 1.7% of the earth's total water and over 30% of the total freshwater supply. Is there enough groundwater to meet human, plant, and aquatic life needs? In many parts of the world, yes; however, with changing demographics and concordant land use and climate change, the distribution and availability of groundwater may not be sustainable. This chapter considers some of the current and past stressors of groundwater by using case examples from around the world. We explore hydrogeologic settings where anthropogenic activity has impaired or has the potential to impair human, plant, and aquatic life. Stressors include well pumping, mining, climate change, chemical use, water law/regulation, and manipulation of surface water. These examples serve to inform those concerned about sustainable management and offer insight into the links between groundwater, climate, and land use.

Keywords: aquifer, discharge, groundwater, recharge, landscape, sustainable

1. Introduction

Human life requires clear, clean, and adequate water to physically survive [1]. Besides direct consumption, water is needed to grow and prepare food. The source of the water for human use comes from the sky and then takes varying paths into the human body, plants, and aquatic life. Fifty percent of the world population obtains their water from groundwater; however, the number is larger when considering that 40% of streamflow is derived from groundwater discharge into a channel during baseflow [1].

Aquatic life is 100% dependent upon water, and when water disappears or changes temperature or becomes contaminated, fish and other organisms die or move or adapt. When a watershed changes vegetation or the amount of impervious surface, the hydrologic pathways shift from infiltration-evapotranspiration to subsurface/interflow to overland runoff. Aquatic life tends to degrade and even disappear as a watershed loses the sustained steady discharge of groundwater into fluvial habitat [2].

Whether human, plant, or aquatic life, groundwater is life-giving; without groundwater our quality of life and the quality of rare plants and aquatic life are less than optimal! Apart from connate water (water held in storage from a different climatic era [3]), groundwater is renewable—the question we address is how we sustain groundwater to meet the current and future demands of human, plant, and aquatic life. The answer is embedded in watershed management; land use decisions in both space and time greatly influence the hydrologic pathways and processes, which also influence human, plant, and aquatic life.

This chapter uses examples of hydrogeologic landscape settings in North America, Europe, Asia, and Africa to illustrate the theoretical movement of water from the sky upon, over, into, and through a watershed. We will address a range of settings and scales to elucidate systems' understanding. It is our hope that this approach will help the reader see critical thresholds that sustain human, plant, and aquatic life in a changing environment. Future groundwater managers will need to grasp the ramifications of their decisions, because like a large ship in the ocean, we can turn or stop the vessel before it may be too late. Well thought-out management decisions about future groundwater supply and demands are needed more now than ever before.

2. Recharge to discharge

What do we mean by *recharge*? It is the water that infiltrates beyond the vadose zone (unsaturated zone) to add to the phreatic zone (zone of saturation); this zone may or may not be an aquifer but part of runoff via interflow and the variable source area that contributes to surface water [2]. Water that *infiltrates* the soil surface does not automatically become groundwater. When plants are present, roots will sequester infiltrated water and pull the water through the plant for physiological needs such as cooling via transpiration. When plants die or go dormant, infiltrated water can move via gravity to the top of the water table or zone of saturation. Temperature can be a factor if water in the vadose zone freezes and becomes immobile until a soil thaw occurs. In the northern hemisphere, the soil thaw occurs in the spring (late March to April). Typically, this is the time of the year water *percolates* through soil pores or fractures to become groundwater. Water that moves in the saturated zone is constrained by the pores (void space) and the pressure or hydraulic head moving groundwater known as *transmission*. The pathway and destination of the groundwater depend on the permeability of the geologic material. Transmissive material is considered an aquifer where water moves relatively quick based on forces of gravity or extraction. Because the earth is not uniform in topography or the size of geologic materials, groundwater will typically move to a *discharge* location over some period. Discharge refers to a point or plane where groundwater is released back to the open free surface. In a natural watershed, these areas of discharge are known as springs, headwater streams, wetlands, ponds, lakes, or even an oasis in the desert [2]. Except for a spring or oasis, it may not be apparent that groundwater is being pushed to the surface. Often, instrumentation is needed to measure groundwater discharge to a surface water body. Yet, where humans have placed a pipe in the ground or water well, discharge occurs through abstraction or pumping for domestic consumption and crop irrigation. The question raised in this chapter is if human interjection in terms of vegetation, land surface management and infrastructure; wells and channel control are changing future sustainability? Human actions have consequences!

3. Hydrogeologic landscapes

The following case examples illustrate both temporal and spatial scales of differing groundwater systems by physical location, topography, geology, vegetation, climate, and land use. We then explain the recharge to discharge story and comment on human alteration and/or benefit, plant floristic value and/or impact, and aquatic life needs and/or impact.

3.1 Large-scale African systems

Table Mountain Group (TMG) aquifer is a regional fractured-rock aquifer located in South Africa where the climate changes with elevation. The aquifer is a major source of water supply for agricultural and urban water requirements in the Western and Eastern Cape Provinces of South Africa [4]. Where the shale layers are not present, groundwater recharge can move deep into the transmissive sedimentary bedrock.

Figure 1 shows a schematic illustration of groundwater recharge and discharge areas and linkages of interaction between surface water and groundwater resources in the TMG aquifer [5]. Groundwater recharge mainly occurs in the higher elevation mountainous terrain areas, while natural discharge occurs in lower elevation valleys and foothills. Nevertheless, shallow groundwater occurs in the alluvial deposits, but downward movement is constrained by shale. The shallow groundwater has a shorter residence time and is not influenced by the more thermally connected mountain recharged water. This water is critical for plant and aquatic life, but during drought conditions, the shallow groundwater can be strained.

The main pathways of natural discharges from the TMG aquifer include 11 thermally heated springs and numerous cold spring discharges up through the quaternary and alluvial sediments providing baseflow to streams and reservoirs, wetlands, and seepage to the ocean [4].

Groundwater discharges naturally and through man-made abstraction via wells. Groundwater is used for portable urban water supply and a variety of agricultural activities. The groundwater is a driving force which sustains human health and regional economy. Natural groundwater discharges from the TMG aquifer contribute to surface water resources in two major ways: firstly as contributions to the flow regime of mountain and foothill streams and rivers and secondly as groundwater contributions to wetlands and other aquatic ecosystems inclusive of marine discharges [4–6]. These natural discharges which take place in different ecotones and scales, as influenced by the subsurface heterogeneity, have an important role for nourishing and sustaining the plant and aquatic life systems in different ways.

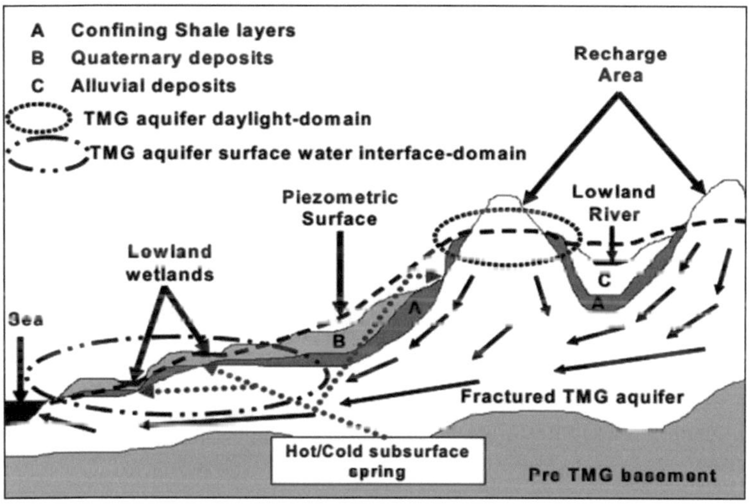

Figure 1.
A schematic illustration of the main groundwater recharge and discharge areas and linkages of interaction between surface water and groundwater resources in the TMG aquifer [5]. Source: [5].

Plants and fish adapt to the temperature and mineral content of the discharging groundwater. As shown in **Figure 1**, protecting the deep groundwater recharge areas is foundational to sustain human, plant, and animal health.

The Stampriet Transboundary Aquifer System (STAS) is shared between Namibia and Botswana, South Africa (**Figure 2**). The largest portions of the aquifer occur in Namibia's arid region which extends to Western Botswana and a small part of South Africa's Northern Cape Province. Auob and Nossob ephemeral rivers constitute the major surface water resources. The groundwater system is composed mainly of the unconfined Kalahari aquifer units overlying the Auob and Nossob confined sandstone aquifers [7].

Research has shown that most of the recharge occurs in the northwestern portion of the watershed in Namibia (**Figure 2**). The recharge typically occurs over a large diffuse area through the unconfined Kalahari formation. Water then preferentially recharges the confined aquifer systems where hydraulic heads and aquifer permeability converge. Several studies strongly suggest that sinkholes and bedrock faults act as the main pathways for preferential recharging of confined aquifers [8–10].

Natural groundwater discharge from the aquifer mainly occurs through evapotranspiration. Groundwater from the aquifer systems evapotranspires due to the aridity of the region. The Auob and Nossob rivers are ephemeral and lack consistent groundwater discharge; only a minimum contribution of groundwater discharge occurs through baseflow into the rivers. Nevertheless, evapotranspiration is also an important process to maintain/sustain the vegetative ecological balance. Given this reality, managers should not expect a robust healthy aquatic life.

Groundwater discharge from the transboundary aquifer also serves basic human needs for drinking and domestic use. Groundwater from the shared aquifer also supports a wide range of industries contributing economic growth and job creation.

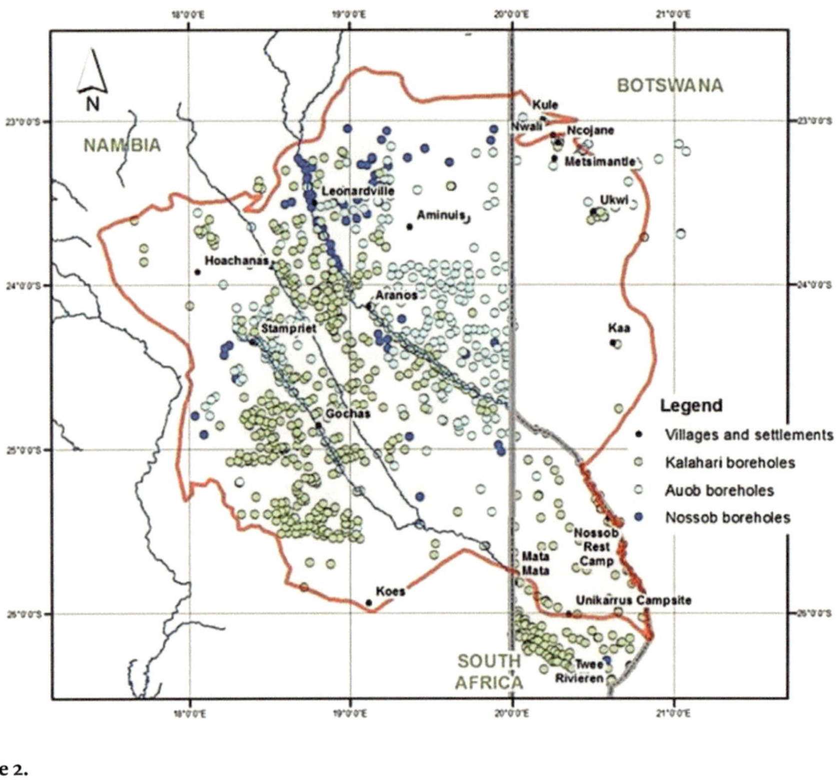

Figure 2.
Stampriet Transboundary Aquifer System and boreholes tapping from the aquifer [7].

Figure 3 shows the relative percentage of land use for each river system [7]. While the scale of groundwater use is different in the three countries, groundwater discharge through abstractions appears to be sustainable.

3.2 North American outwash sandplains: wildlife and irrigated row crops

The Anoka Sand Plain in Minnesota, USA (**Figure 4a**), and the Central Sands region of Wisconsin (**Figure 5**) are formed by rapid glacial melting which allowed meltwater to carry fine sediment south toward the Mississippi River; however, coarse-grain sediment was dropped out quickly to form large flat areas composed mostly of sand with gravel.

Because dense compacted till was laid down by ice advances from Canada, infiltrating water can fill up the surficial sand and gravel aquifer but not move laterally unless a stream, lake, or wetland exists. Stream gradients in these regions are flat because the landscape is flat. In low-lying terrain, the groundwater manifests itself as large wetland complexes, whereas higher ground contains oak-prairie savannahs and crops. Some of these areas are protected by federal and state wildlife legislation [12], but most of the land is in some form of agricultural management. Because the soil has a high sand content, summer evapotranspiration can quickly dry up the upper topsoil, such that only deep-rooted plants survive the warm summer temperatures. High-value crops, such as potato and other vegetable crops, require irrigation to optimize plant vigor and specialty crop quality. In other locations, drainage via ditches is required to prevent crop loss. Some wetlands have been drained to grow grass, known as sod. Because sod is a high-value crop, pumps and lift stations are needed to prevent crop loss due to saturation. Because outwash regions are very flat, groundwater moves slowly unless an artificial gradient is created by ditches and pumps. There is an ongoing battle between nature and human development; the subdevelopment of homes, streets, parking lots, and shopping malls leads to urban runoff and stress upon the plant and animal (wildlife) ecological equilibrium.

In the Central Sands region, intensive agricultural production has led to elevated nitrate-nitrogen concentrations which have threatened domestic drinking water users [14]. Aquatic life does not thrive well in flat channel gradients and sandy substrate; the flat terrain does not allow an adequate cold-water fishery, even though water temperatures in some channels are controlled by groundwater. Nevertheless, amphibian, mammal, and bird wildlife are abundant in wildland areas. Without governmental protection, these areas are at risk of losing their biodiversity [15].

3.3 Incipient karst: non-laminar groundwater flow

Well-developed karst features are present in several parts of the world, most notably Croatia in Europe and Kentucky, USA. Incipient karst differs from developed karst because solution enlargement of fractures has not created caves. The thickness

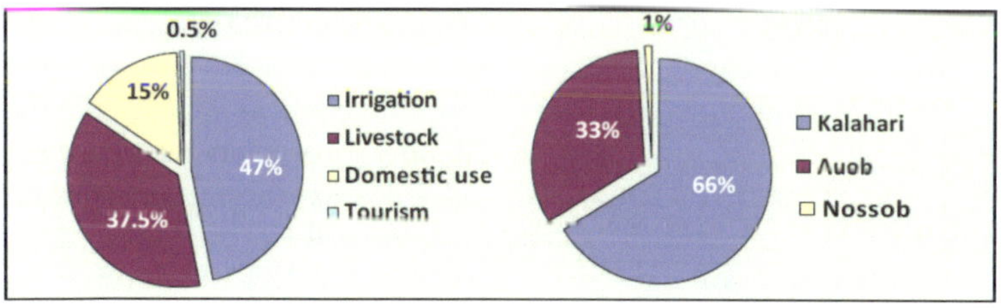

Figure 3.
Groundwater use and abstraction per aquifer type in the Stampriet Transboundary Aquifer System. Source: [7].

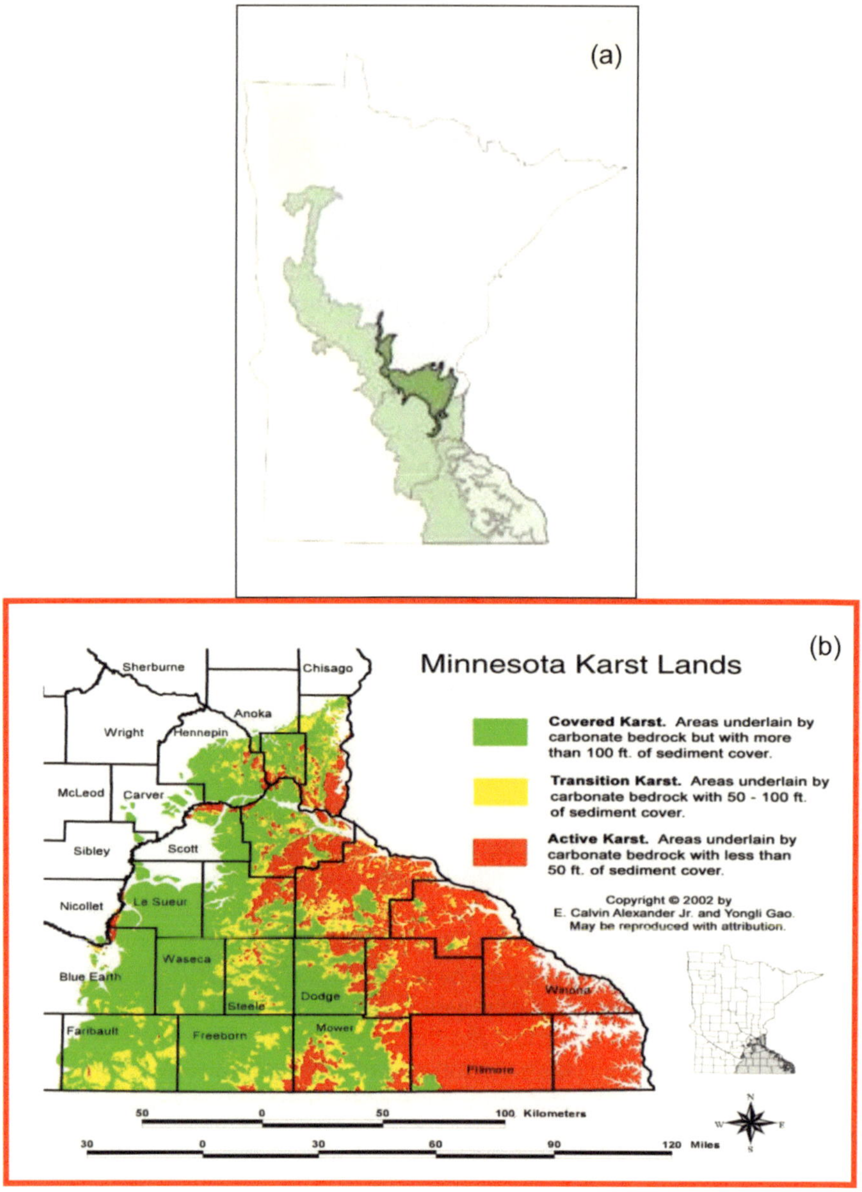

Figure 4.
(a) Illustration of the boundary of the Anoka Sand Plain highlighted in central Minnesota—darken area with a border. (b) Karst areas southeast of the Anoka Sand Plain in Minnesota [11]. Source: Environmental Trust Fund.

of soil cover somewhat defines the boundary between incipient and developed karst (**Figure 4b**). Developed karst is dominated by sinkholes, underground cave streams, and point-source springs. Incipient or immature karst aquifers can have rapid water movement but may not have a landscape dotted with sinkholes and springs where cave streams resurge. Soil pipes are present, but they open and close quickly depending on cohesive soil bridging over bedrock. Subsurface erosion occurs through rock fractures in both carbonate and sandstone rock if the overlying soil is dominated by silt. The silty soil will bridge above a cavity until the soil-bearing strength is exceeded or triggered by changes in soil moisture or land use. The lack of abundant sinkholes makes land use development challenging because short of ground-penetrating radar or other geophysical measurement, there is no way to be certain a structure will not be swallowed by catastrophic collapse at some future date [16].

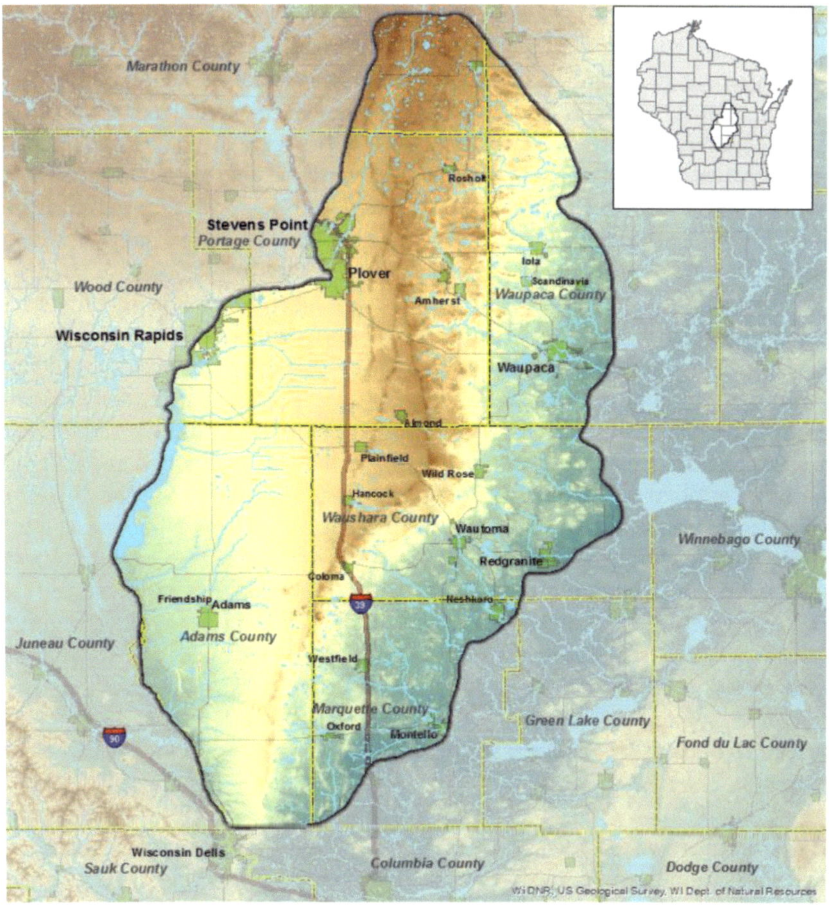

Figure 5.
Location of the Central Sands region in Wisconsin [13]. Source: WiDNR webpage.

With a changing climate, the risk of weak carbonic acid water in contact with limestone/dolomite fractures may drive fracture-opening enlargement, increasing the uncertainty of making sustainable groundwater management decisions. If all fractures below the landscape have a similar size and connectivity, Darcy's law and laminar flow can be assumed and modeled; however, nature's enlargement processes are not uniform but chaotic. This chaos leads to highly unpredictable flow paths and pollutant transport in relatively short time scales: on the order of minutes-to-days compared to months-to-years for Darcian flow. Hydrocarbons, pesticides, nutrients, and bacteria have been observed to move from the surface/near surface into shallow domestic wells and spring-fed streams over short time scales [17]. Future groundwater management will require a field monitoring effort that is more rigorous than sand and gravel systems. Tracking water and solute movement will require a monitoring system like those used in urban watersheds where water movement responds quickly to new precipitation. The largely unknown factor is storage; sustainability will depend upon creating watershed storage to buffer adverse ecosystem service change.

3.4 Midwest US Corn Belt: shallow groundwater

Midwestern states of the USA produce the largest amount of corn grain on the planet. To achieve the massive amount of grain production, the land must be intensively managed to optimize crop growth. Large equipment for planting and harvesting, state-of-the-art agronomic practices from seed, chemical inputs to the grain storage, and marketing drive the Midwestern US economy. The landscape was once a vast sea of deep-rooted prairie grass and wetlands which helped form black fertile carbon-rich soil. In many locations soil wetness created uncertainty in crop management. To address this problem, wide-scale ditch drainage began over 100 years ago to optimize plant growth in wetland environments. Today fewer ditches are dug, but the use of plastic corrugated and perforated pipe that is placed into the soil with laser accuracy is a booming business. This land use practice helps remove excess water in the upper meter of cohesive soil; typically, sandy soils do not use subsurface pipe to improve soil aeration. The cohesive soil acts like incipient karst allowing water in the soils to move rapidly into the pipe because of fractured soil structure. In some ways the subsurface pipe functions like an urban environment producing pipe flow that transfers water to streams and ditches during and after a rainfall event. This change in hydrologic connectivity has caused downstream channels to enlarge over time leading to unstable banks and beds. Further, the chemicals applied to farm fields can move downward into the pipe and cause eutrophication of downgradient surface water. Nitrate-nitrogen has increased with increased placement of pipe; this has, in turn, led to Gulf of Mexico hypoxia [18]. To find a sustainable solution to this problem, water managers will need to find ways to hold water back and treat polluted runoff. Building soil health, regulating pipe discharge, and using bioreactors and saturated buffers are tools to be examined to minimize sediment and nutrient problems to downstream waterbodies.

3.5 Eastern Himalayas: Mizoram springs—will they be sustainable in the future?

Located in the northeastern states of India, east of Bangladesh (**Figure 6**), this landscape is extremely steep with long narrow valleys. The soils are very thin over shale or sandstone. The vegetation is thick and lush given the monsoon rainfall for 5 months of the year. Even though the landscape is covered in perennial vegetation, the water will only infiltrate centimeters before it enters the sandstone or converges as a headwater stream running off to a river some 2500+ meters below.

People depend on water infiltrating the sandstone and then resurging downgradient as a spring for human use during the non-monsoon season. If the landscape is disturbed by slash and burn agriculture, then less aquifer recharge occurs. The solution requires more sustainable agricultural land use and strategically planned capture of monsoon runoff water. The magnitude and intensity of rainfall during the monsoon season may be shifting in a way that is limiting aquifer recharge to occur. If more water is running off the landscape compared to past decades, then land use must adjust to hold back runoff. This not only means less bare soil but improved soil infiltration and aggregate stability. Topsoil must be highly valued and managed to optimize soil health [19]. In selected ravines, a portion of overland runoff should be laterally diverted wherever a slope break occurs; this practice can provide focused recharge into sandstone aquifers to augment water storage and availability during the non-monsoon seasons.

3.6 Eastern front of Rocky Mountains: alpine to semiarid water law

Located along the eastern front of the Rocky Mountains in Montana (**Figure 7**), this landscape is weathered due to wind and water erosion. At high elevations (4000 meters+), temperatures remain cool to cold, so vegetative growth is stunted [20].

The region is managed to capture and hold snow for summer water supply to the dry eastern plains. Soils are loamy over metamorphic bedrock, so there is no deep

Figure 6.
Location of Mizoram in Asia. Source: Google Images.

recharge, just overland and interflow to channels that are formed by snowmelt. Near the toe-slope of the range, water can spill into meadows where more organic-rich loamy soils are mixed with stratified layers of sand and gravel and cobble, allowing snowmelt water to infiltrate and recharge shallow aquifers. Because of a less steep gradient in places, landowners have dug ditches to divert streams into pasture lands. Both natural and diverted waters resurge in wetlands or springs depending on the geologic constraints. However, the water that does not evapotranspire will move out across the land only to infiltrate into the sediment/soil depending on the nature of the geologic material. This occurs because the source of the water is still snowmelt from high-alpine elevations. Though the water has been geochemically transformed by passing through rock and sediment, it is not groundwater in the sense of an aquifer that provides decadal storage. Further, because this water does not always remain in the stream channel as it flows east downstream, users do not receive the benefits of the alpine water because the water will seep through the channel bed into an aquifer. The new climatic reality demands that water managers examine law and policy to find a sustainable way forward [21].

3.7 The North American Great Lakes and groundwater

The Great Lakes in North America provide unique freshwater resources for humans, plants, and animals. Many small tributaries drain directly to a large lake, but in some geologic settings, groundwater discharges into Lake Michigan through

Figure 7.
The location and scenery of the Bob Marshall Wilderness. Source: gravel.org.

sedimentary rock and breach ridge sands. In Door County, Wisconsin (**Figure 8**), coastal springs and wetlands provide ecotones for rare species like the Hine's emerald dragonfly and unique orchids, such as lady's slipper. The Ridges Sanctuary was created in 1937 to sustain plant and wildlife in an area that was rapidly developing to accommodate tourist demands. Near-shore areas were being developed for lodging and food and drinking establishments. Given the high tourist value of Door County, Wisconsin, today, it was possible for people to love the place to death. Specifically, if groundwater recharge areas were paved and rain water redirected to streams, ecosystem services would have been lost. Fortunately, visionaries like Albert Fuller sounded the alarm to the general public and raised awareness to preserve an 18-Ha parcel of land from future development. Over time, studies in Door County, Wisconsin, have provided more information about the natural resources and the need for local government to place restrictions on land development. Groundwater flows from higher ground underlain by dolomite toward Lake Michigan at a rate approaching a cm/second [22]. The large lake waves can push water and sediment back unto the land; over time this process has created sand dunes. At the Ridges Sanctuary, there are a series of dunes with swales, between the dunes are wetlands that provide critical habitat for plants and animals.

Along the north shore of Lake Superior, the geology is metamorphic and gives rise to steep gradients within a kilometer of the shoreline, which differs from the relatively flat shoreline of Door County. The water that infiltrates the shallow soil

Figure 8.
The location of Door County in Wisconsin, an aerial view of the Ridges adjacent to Lake Michigan and vegetation. Source: map.co.door.wi.us., mnnps.org.

and moves downgradient toward Lake Superior resurges where the soil becomes too thin over the bedrock. Glacial Lake Duluth left behind linear zones of sediment: some beach sands and other lacustrine silts and clays. In Amity Creek before branches converge, the valley slope flattens, and alluvial material creates an active flood plain over bedrock. In June of 2012, after a year of data collection from the stream and alluvial aquifer, a large magnitude storm event dropped over 15 cm of rainfall in a half-a-day. This event not only flushed channels; it displaced preexisting snowmelt water contained in the riparian aquifer [23]. We have further noted a complete groundwater flushing from the same storm event in the Cross River watershed [24]. The data gathered from these two watersheds indicate a lack of resilience to climate change. To maintain sustainability for the high-valued tourist region, infrastructure development along the north shore of Lake Superior must be constrained to prevent the loss of ecosystem services. All levels of government will need to agree on the vulnerability of the region.

In the Nemadji River basin, lake sediment dominates the movement of groundwater (**Figure 9**). The west end of Lake Superior was formed by beach ridges laid upon coarse till, whereas the central part of the valley is composed of loose lacustrine silts and clays that settled when glacial Lake Duluth drained to the east. This sediment deposition pattern creates a unique groundwater flow system. Water recharges rapidly through sand and gravel in the headwaters but then builds up pressure as it tries to find a discharge path into the Nemadji River. Because there is over 25 meters head drop from the headwaters to the main valley, the valley walls are under hydrostatic pressure and ooze water through the lacustrine silts and clays. Bank and bluff geotechnical failure are a natural phenomenon that creates a continuous turbid water clarity in both discharging groundwater and surface water [25].

The loss of large perennial trees and conversion of land to managed grass crops altered the hydrologic evapotranspiration regime which, in turn, increased runoff and concordant river sediment regime. The Nemadji River system is the most productive trout fishery in Western Lake Superior, but channel bed downcutting has the potential to create fish barriers and block trout migration. Long-term aquatic

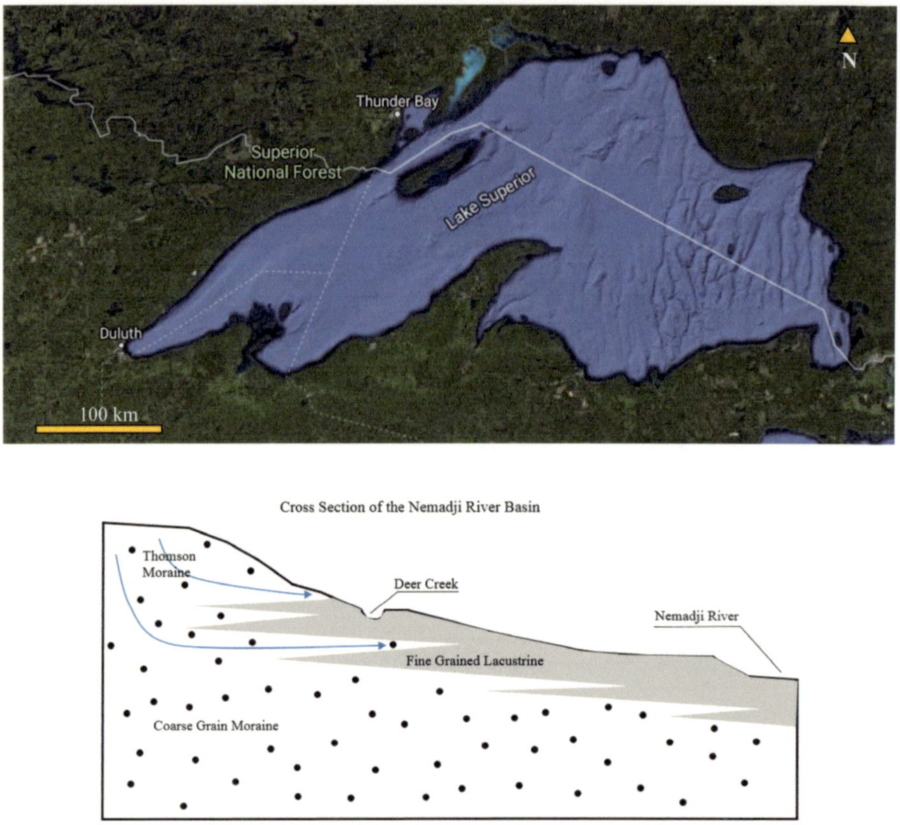

Figure 9.
The location of Lake Duluth on the west end of Lake Superior and illustration of groundwater flow paths southeast of Lake Duluth, Minnesota, in the Nemadji basin [25]. Source: Google Images and Magner PowerPoint.

life sustainability will depend on land use management and then enhances evapotranspiration, reduces runoff, channels enlargement, and allows for fish passage culverts.

3.8 Trapped groundwater: mining the Buffalo aquifer

Connate water is water that is not actively part of the water cycle but groundwater contained or trapped in the earth's crust from some previous time period. In the northern latitudes of North America, Europe, and Asia, this may be frozen groundwater or water left behind in a buried aquifer when glaciers retreated toward the Arctic region.

Climate change in the Arctic region may be liberating frozen groundwater today that has been contained due to a lack of any hydraulic head to move the water toward discharge.

Magner and others [26] used isotopes to estimate the age of water contained in a buried sand and gravel aquifer embedded in the lake clays left behind by Glacial Lake Agassiz. The City of Moorhead, Minnesota (**Figure 10**), needed to expand their water supply and began pumping tests to determine the sustainability of the Buffalo aquifer. The results suggested the high-capacity water abstraction would lead to groundwater mining; thus, the city focused their water supply efforts toward the Red River of the North. Nevertheless, single-family homes with small domestic water demand could pull water from buried sand and gravel. Over a long

Figure 10.
The location of Moorhead, Minnesota, and domestic water. Source: claycountymn.gov and valleynewsalive.com.

period of time, very slow-moving groundwater would likely replenish the buried aquifer water volume; however, this may take millennia. The city of Moorhead, Minnesota, made the right and sustainable decision; however, in California, parts of the Central Valley are sinking as both farms and cities pump harder and drill deeper wells to extract groundwater. The California groundwater is estimated to be 15,000–20,000 years old [27]. This is perhaps the best example of a truly unsustainable groundwater use in the world.

3.9 Protecting rare plants

Groundwater contained in a limestone or dolomite aquifer that discharges into a major river basin provides a calcium-, magnesium-, and bicarbonate-rich water that drives rare calciphile plant occurrence, converging streamlines of groundwater flow into valley fen. Shallow flow paths have short travel times based on anthropogenic chloride and sodium typically not found in the deeper flow paths of a carbonate aquifer. Thick calcite accumulations occur in the root zone at the water table.

There is a mixture of upwelling groundwater and water near the surface that can mix and then flow downslope from higher elevations into the fen or toward spring-fed lakes or directly to the river (**Figure 11**). Shallow groundwater decreases downgradient in the calcareous fen as older groundwater pushes up to discharge [28]. Komor [28] believes that encroachment of reed grasses and other invasive species into the calcareous fen may reflect human-caused disturbances in the valley. The land use in this riparian area requires special protection to limit invasive species and preserve the unique plant life. The Minnesota Department of Natural Resources has enacted rules to sustain and preserve calcareous fens in Minnesota.

3.10 Kura River basin: sustainable sturgeon

The Kura River basin (**Figure 12**) has been poorly managed over time [29]. Climate change has shifted the timing of melting snow and the baseflow in the Kura River. But the over extraction of groundwater for industrial and agricultural use has led to an adversely impacted fish habitat, and sturgeon reproduction has diminished over the past half century.

Environmental flows are needed to find a sustainable solution to meet the demands of an important part of the Caspian Sea economy, namely, black caviar. A key factor is the minimum baseflow required to maintain sediment transport and fish habitat. Pools and riffles are fluvial features formed by the transport of coarse-grained sediment: sand and gravel. If channel-forming flows are disrupted with too little runoff, pools will fill with aggraded sand. This problem was further exasperated by industrial harvesting river beds to obtain well-sorted aggregate for road construction. The damage primarily occurred under the soviet era management of the region. Today attempts are made to bring the watershed back to the environmental flows needed to support a sturgeon fishery. This is a recognition by the Azerbaijan officials that valley groundwater is not unlimited and that competing demands of industrial and agricultural interests must include riverine habitat to support fish.

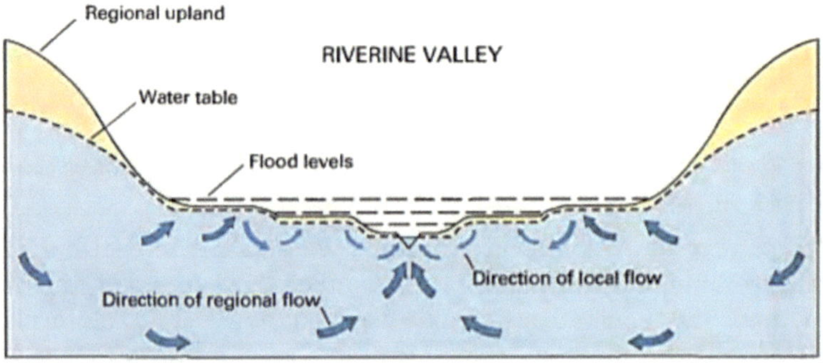

Figure 11.
Illustration of large valley groundwater flow paths like what occurs in the Lower Minnesota River Valley [1]. Source: USGS webpage.

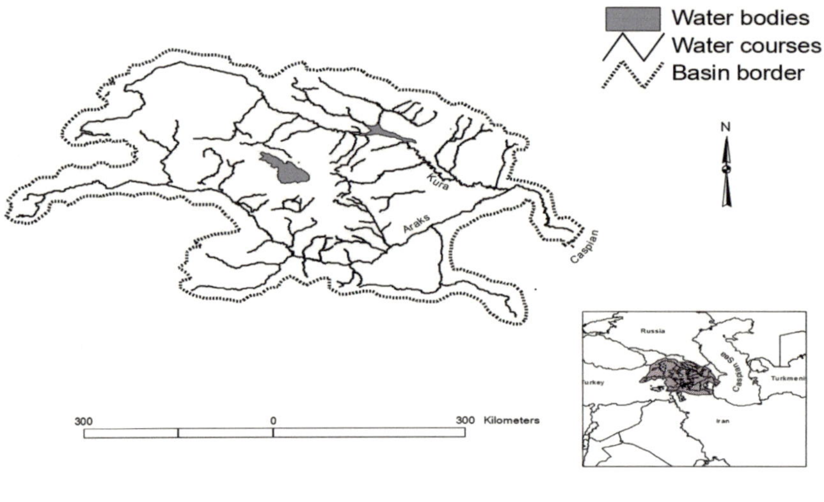

Figure 12.
Location of the Kura River basin west of the Caspian Sea [29]. Source: Abbasov PowerPoint.

4. Conclusion

This chapter has provided some basic information about the relationship between groundwater sustainability and the geology, climate, and land use of various hydrogeologic settings on four continents. This chapter draws from the life experience of the authors and presents case example stories in a manner that hopefully allows nontechnical readers to understand the interface of anthropogenic activity and the sustainability of humans, plants, and animals.

Acknowledgements

We acknowledge our past undergraduate and graduate students who spent many long hours in the field collecting data, analyzing data, and building a compelling argument to define hydrologic pathways and processes.

Author details

Joe Magner[1] and Modreck Gomo[2]*

1 Department of Bioproducts and Biosystems Engineering, University of Minnesota, St. Paul, MN, USA

2 Faculty of Natural and Agricultural Sciences, Institute for Groundwater Studies, University of the Free State, Bloemfontein, South Africa

*Address all correspondence to: gomom@ufs.ac.za

References

[1] Winter T, Harvey J, Franke O, Alley M. Ground Water and Surface Water A Single Resource. U.S. Geological Survey, Circular 1139. Denver, CO: 1998

[2] Brooks K, Ffolliott P, Magner J. Hydrology and the Management of Watersheds. 4th ed. Hoboken, NJ: Wiley-Blackwell; 2013. 533 p

[3] Fetter C. Applied Hydrogeology. 2nd ed. Columbus, OH: Merrill; 1988. 80 p

[4] Duah A. Sustainable Utilisation of Table Mountain Group Aquifers [PhD thesis]. University of the Western Cape, South Africa. 2010. Available from: http://etd.uwc.ac.za/xmlui/bitstream/handle/11394/2981/Duah_PHD_2010.pdf;sequence=1

[5] Roets W, Xu Y, Raitt L, Brendonck L. Groundwater discharges to aquatic ecosystems associated with the Table Mountain Group (TMG) aquifer: A conceptual model. Water South Africa. 2008;**34**:77-88

[6] Roets W, Xu Y, Raitt L, El-Kahloun M, Meire P, Calitz F, et al. Determining discharges from the Table Mountain Group (TMG) aquifer to wetlands in the Southern Cape, South Africa. Hydrobiologia. 2008;**607**:175-186. DOI: 10.1007/s10750-008-9389-x

[7] UNESCO-IHP. Stampriet Transboundary Aquifer System Assessment—Technical Report. Governance of Groundwater Resources in Transboundary Aquifers (GGRETA)—Phase. Vol. 1. Paris: UNESCO; 2016

[8] van Wyk E. Southern African Pre-Cretaceous deep groundwater flow regimes: Evidence and drivers. Golder Associates Africa Report (Pty) Ltd, Midrand, South Africa. 2014

[9] Kirchner J, Tredoux G. Applying environmental isotopes to a hydrogeological model of the Stampriet Artesian Basin Project RAF 8/029. 2002. 62 p

[10] Tredoux G. Die korrelasie tussen waterkwaliteit en geologiese formasie in SWA: 'n Geohidrochemiese ondersoek van die Artesian Kom Stampriet. Bloemfontein: UOVS; 1981. 297 p

[11] Alexander EC Jr, Gao Y, Green JA. Minnesota Karst Lands. 2006. Available from: https://www.esci.umn.edu/sites/www.esci.umn.edu/files/user/user174/MN Karstlands 2006.jpg

[12] US Fish and Wildlife Refuge. Available from: https://www.fws.gov/refuge/Sherburne/about.html

[13] Wisconsin Department of Natural Resources (WiDNR). Web-access. 2018

[14] Kraft G, Clancy K, Mechenich D, Haucke J. Irrigation effects in the Northern Lake States: Wisconsin central sands revisited ground water. 2012;**50**(2):308-318. DOI: 10.1111/j.1745-6584.2011.00836.x

[15] Meretsky V, Fischman R, Karr J, Ashe D, Scott M, Reed N, et al. New directions in conservation for the national wildlife refuge system. Bioscience. 2006;**56**:135-143

[16] Magner J, Book P, Alexander EC Jr. A waste treatment/disposal site evaluation process for areas underlain by carbonate aquifers. Ground Water Monitoring Review. 1986;**6**:117-121

[17] Crawford K, Lee T. Using nitrate, chloride, sodium, and sulfate to calculate groundwater age. In: the Proceedings of the 14th Sinkhole Conference. 2015. pp. 43-52

[18] Magner J, Payne G, Steffen L. Drainage effects on stream nitrate-N and hydrology in south-central Minnesota (USA). Environmental Monitoring and Assessment. 2004;**91**:183-198

[19] Zonunsanga R, Magner J. Hydrological engineering for sustainable shifting agriculture in the Eastern tropical Himalayas: A conceptual discussion. Science Vision. 2017;**17**:55-60

[20] Bob Marshall Wilderness. 2018. Available from: https://www.wilderness.net/NWPS/wildView?WID=64

[21] Grafton R. Global insights into water resources, climate change and governance. Nature Climate Change. 2013;**3**:315-321

[22] Cobb M, Bradbury K. Best Management Practices to Protect Groundwater at Hine's Emerald Dragonfly Larval Sites in Door County, Wisconsin: Cooperative Agreement No. F12AC00153 Between the U.S. Fish and Wildlife Service and The Ridges Sanctuary. The Ridges Sanctuary. 2013. Available from: https://www.fws.gov/midwest/endangered/insects/hed/pdf/HEDBMPFinalReportFeb2013.pdf

[23] Jasperson J, Gran K, Magner J. Seasonal and flood-induced variations in groundwater–surface water exchange in a northern coldwater fishery. Journal of the American Water Resources Association. 2018;**54**:1109-1126

[24] Magner J, Zhang L. Cross river watershed hydrologic adjustment pre- & post-June 2012 mega-storm. Journal of Environmental Science and Engineering B. 2014;**3**:133-141

[25] Magner J, Brooks K. Predicting stream channel erosion in the lacustrine core of the upper Nemadji River, Minnesota (USA) using stream geomorphology metrics. Environmental Geology. 2008;**54**:1423-1434

[26] Magner J, Regan C, Trojan M. Regional and local flow systems in the Buffalo River watershed, Northwestern Minnesota. Hydrologic Science and Technology. 2001;**17**:237-246

[27] U.S. Department of the Interior | U.S. Geological Survey. 2017. Available from: https://ca.water.usgs.gov/land_subsidence/index.html

[28] Komor S. Geochemistry and hydrology of a calcareous fen within the Savage Fen wetlands complex, Minnesota, USA. Geochimica et Cosmochimica Acta. 1994;**58**:3353-3367

[29] Abbasov R, Smakhtin V. Introducing environmental thresholds into water withdrawal management of mountain streams in the Kura River basin, Azerbaijan. Hydrological Sciences Journal/Journal des Sciences Hydrologiques. 2009;**54**:1068-1107. DOI: 10.1623/hysj.54.6.1068

Chapter 5

Mining of Minerals and Groundwater in India

Abhay Kumar Soni

Abstract

Mining of minerals is essential for our day-to-day life so is the groundwater. Mother Earth is the custodian of these two essential commodities, and both are part and parcel of sustainable living for human beings. This chapter of book focuses on the need, quantity, quality, and management of groundwater encountered in mines, from where extraction of minerals takes place. By understanding interrelationship between groundwater hydrology and mining, the basic objective of sustainability, that is, conserving for future generations with particular reference to the mines, has been addressed. Such scientific approach makes the mine planning easier, ensures better water management, and solves water scarcity as well as security problems in the vicinity of mining areas.

Keywords: surface mining, underground mining, impact of mining on groundwater, quantitative estimation in a pit, statutory compliance, groundwater in hard rocks

1. Introduction

Minerals and their exploitation had been carried out since centuries by two major methods, namely, surface mining methods and underground mining methods. In both these methods, groundwater role is important as well as advocated because mining has influence on hydrology. While permitting mining, the disturbance to the hydrological regime should be minimum or as less as possible.

It is beyond doubt that for food security, human health, energy, and ecosystem, groundwater is absolutely important for the entire world [1].

This groundwater is continually being put under increasing stress because of the industrialization, growing needs of the population, and its improper use as a resource. Its mismanagement has led to uncalled water scarcity in present time and also threatened us with water pollution problems. Groundwater science and its accurate estimation for the mining areas are a bit cumbersome because the dynamics of groundwater keep on changing as excavation size is changed. Therefore, the role of groundwater in mining of minerals assumes special emphasis which is analyzed and discussed as a separate chapter in this book. This knowledge, though very exhaustive, will be certainly helpful for the mining areas and mega-sized mining/mineral sector in improving the quality of human life.

While dealing with water problems of mines, three keywords must always be remembered as they are extremely important, namely, mine water (MW), groundwater (GW), and surface water (SW). Our focus in this chapter has been kept on mine water and its analysis with respect to the two principal methods of mining, that is, opencast mine and underground mine only. Besides these principal methods, other methods are not covered, though other novel methods and technologies of mining do exist, for example, solution method, mechanical method, aqueous extraction methods (hydraulic mining), etc.

The MW analysis automatically covers SW and GW, as mine water is either/or a combination of both for all mineral types (categorized as fuel minerals, that is, coal and lignite; metallic minerals, that is, iron ore, bauxite, etc.; nonmetallic minerals, that is, limestone, dolomite, etc.; and minor minerals, that is, sand, building materials, etc.) and their extraction from earth called "mining of minerals."

In this introductory paragraph, it is apt to highlight some basic points of groundwater to deal mining of minerals, scientifically. "Groundwater" in surface mines is found below the water table and covered by a layer of soil and/or rock. Groundwater is always present at below ground level and indirectly available at the mine pit as "base flow." It gets intercepted while excavating mineral(s) in open mines. Availability of groundwater in open-pit mines and underground mine workings has number of differing dimensions of basic hydrology influenced by site-specific geology. Thus, it requires basic knowledge of water flow and water movement (Darcy's law). Groundwater of mining area occurs in aquifers which are of different categories, namely, unconfined aquifers, semi-confined aquifers, and confined aquifers. The groundwater is contained either in the rock pore spaces or rock fractures/cracks depending on the rock types. Compared to the surface water, it is generally considered to be less easily contaminated, but this does not infer that groundwater is safe from pollution perspective. The groundwater can become contaminated where polluted runoff seeps through the ground to the water table or flows down through fractures or cracks in bedrock (seepages). Wherever surface water bodies are fed from groundwater sources, water contamination may be present in both, though isolated by ground cover. In addition, groundwater often contains dissolved minerals as a result of prolonged contact with rocks containing minerals of different types and varieties which can alter its quality, for example, the presence of arsenic, nitrates, and fluoride [2–6] in aquifers has been reported, and this is an indication for this. The depth of the groundwater at which it is present in and around the mine area is a one major point of observation as well as concern for mine water-related issues.

To understand groundwater-related problems of mine, hydrological and geological setup of the area is first studied. With reference to any mine or the mining area, hydrogeological setup encompasses aquifer characteristics, that is, nature, type, parameters, etc.; all local and regional geological details; and plans for mining and total picture of hydrology, drainage, discharge, etc. The approach for scientific investigation, to search solution, usually includes field monitoring (pre-monsoon and post-monsoon monitoring), instrumental survey (e.g., Resistivity Image Profiling Survey and GPR Survey, etc.), groundwater modeling, and mine planning, that is, drainage, dewatering, etc. Mine being a production enterprise (unit) requires its assessment from industrial perspective; hence, this chapter makes no pretense of neither mining engineering nor of hydrology but explains to the reader the interrelation of mining process with water in general and groundwater in particular.

Here, it is equally important to describe briefly what *new insights* the work has added in terms of knowledge on top of the existing knowledge. In general, mining of minerals and groundwater pertains to the open-pit mining of minerals. Technical literature is also vogue in terms of analysis and with particular reference to the surface mines only. Very little had been dealt about different aspects of underground versus groundwater in mining science. But in this book chapter, both "underground mining" and "open-pit mining" knowledge have been dealt together forming a consolidated base and considering that groundwater is equally important for operative underground mines. Such attempt will provide total and at-a-place look to the reader. Not only this, but it will also enhance further scope of knowledge development, to be done by other researchers, in underground mines/mining and other excavations, which are less researched. Site-specific and typical field conditions will certainly add further to the existing groundwater knowledge base and make underground excavations further safe as well as productive.

2. Water in mines: pollution, discharge, control, and treatment

Water in mines, that is, "mine water," usually refers to the water contained in the mined-out open area or dug-out area generated as a result of mining of mineral. This excavated area is in open-pit form and contains surface water as well as groundwater. In the case of underground mine, the water encountered is principally groundwater. To address technical issues, it is always better to consider them as two entities, that is, surface water and groundwater (**Figure 1**), though difficult, to categorize in the case of surface mine.

The principal source of mine water is the "rainfall," and other possible sources could be enumerated as:

1. Intersection of water table during mining

2. Seepage water

    a. Nearby major water bodies in and around the mining area

    b. Nearby mine workings, may be surface or underground

3. Incessant rainfall/heavy downpour

Mine water, a valuable commodity, is also a form of industrial wastewater (effluent) which can be a disaster in mining areas or a boon to ease the water scarcity problem locally. Both SW and GW are considered at all stages of the mining operation starting from planning to extraction to restoration stage. Different aspects

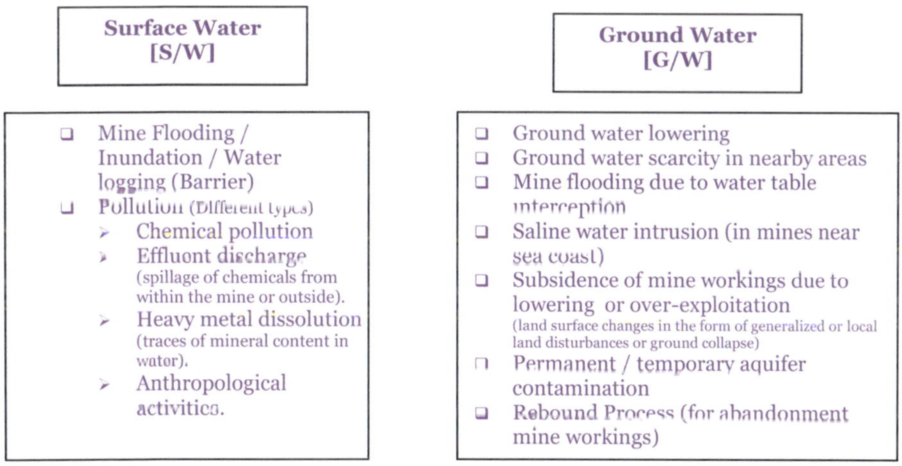

Figure 1.
Problem and issues of water in mines.

of water covered include collection and handling of hydrological data, control of runoff, magnitude of water, diversion of water channels in mines, (if any), erosion and sediment control, dewatering, different water pollution forms as observed in mines, and water management. As said above that the interrelationship of water and mine is complex and far reaching, the solution should be practical to ensure the efficient running of mining operation while adequately protecting the environment.

2.1 Pollution

It has been observed that water pollution in mines is common and well described but their scientific importance is often ignored while managing the mine production. The reasons for this are enumerable. It is desirable that every mine's water, if present, is turned into a useful asset. In some situations water management at mine is neither environmentally friendly nor comparatively easier to manage, for example, acid mine/rock drainage (AMD/ARD), and its management is a costly affair compared to higher TDS water in a limestone quarry. Therefore, sincere attempts have to be made to ensure that AMD pollution and high TDS hard water are treated properly. Similarly, elemental concentration must be checked within permissible limit.

When water comes in contact with exposed mineral at the mine (either at pit or in underground), the potential for water contamination increases manifold. In order to reduce and minimize the water pollution requiring treatment, various control techniques are available. On case-to-case basis and looking at the type of mineral mined, the mine water pollution are dealt for different solutions, for example, heavy metal contamination into water, thereby raising pollution levels are quite frequent in the case of metallic mines.

Mine water control techniques and their selection strategies are cost based and site-specific. It should be carefully selected to prevent the release of contaminated water into the environment. From area to area, one or combination of more than one method may be applied for the pollution control. With high rates of precipitation in an area, significant emphasis must be placed on drainage and its combinations in varying topographies, whereas the mine environment in arid regions with little water availability must choose *water recycling* as the technique of mine water containment for pollution abatement. If pollution has to be controlled and contained in the mining areas alone, the mine water discharge must be routed effectively. This will make the water pollution management more cost-effective. Judicious utilization of water for the appropriate purpose and water conservation for the future need should be implemented into practice as per the law of land. In India, for prevention of pollution due to mine water, the principal act is Water (Prevention and Control of Pollution) Act, 1974 (amended in 1988 and 2003). This act, despite the prevention and control of water pollution, also guides for the maintaining or restoring of wholesomeness of water.

The topic of pollution is so vast and varied that its description in limited pages is beyond the scope of this chapter. Therefore, readers are advised to consult specific literature related with the problem.

2.2 Discharge

Water discharge from a mine is often controlled by effective drainage around. In India, water discharged from mines are governed by general discharge standards/limits framed by the Ministry of Environment, Forest and Climate Change (MOEFCC) Govt. of India [7]. These effluent discharge standards of India containing about 33 pollutant parameters are framed under the Environment (Protection) Rule, 1986 (under Schedule VI).

Discharge of mine water into natural drainage system without any treatment is also an issue to be reckoned in mining geohydrology. In order to avoid the degradation of downstream water channels by excessive suspended sediments from mine, all runoff leaving the mining area should be routed through "sedimentation pond" where the suspended solid can be reduced to acceptable limits. Factors to be considered during design and construction of sedimentation pond include hydrology, its location in mine, construction material and its cost, maintenance/cleanout operations, and applicable legislative requirement [8].

In mining areas, either dendritic pattern or parallel drainage pattern is often present (**Figure 2**).

Intercepting and diverting surface water (rainwater, runoff water, stream water, snowmelt water, etc.) from entering the mine site are the first step to tackle water accumulation in a pit. Since surface mining causes land disturbance including the removal of vegetation, increased runoff, erosion, and sediment, every attempt should be made to control the mine water discharge. Proper relief and gradient together with adequate slope design are helpful in capturing drainage water which can control runoff and erosion of soil as well as sediments. Topography and watershed details of the project area are equally important from drainage and discharge angle. For the mine water discharge, the knowledge of flow direction together with reduced level (RL/MRL) detail helps in planning. Small seasonal nallahs/streams with first-, second-, and third-order drainage pattern are observed in the mining region. Drainage map of the studied mine area is generally drawn covering core zone or CZ (5 km radius) and buffer zone (BZ) of the mine lease (10 km radius). Together with drainage and the watershed area details, an assessment about the seepage from the pit, mine dumps and tailing dams, etc. in nearby area can be made. Such analysis provides the basis for delineating control measures of seepage flow and water management.

In mine-related studies, generally the term *watershed* [classified as first-, second-, third-, or higher order watershed or else they could also be delineated as micro (3000–5000 ha area), macro (>5000 ha area), or mini watershed (<3000 ha area)

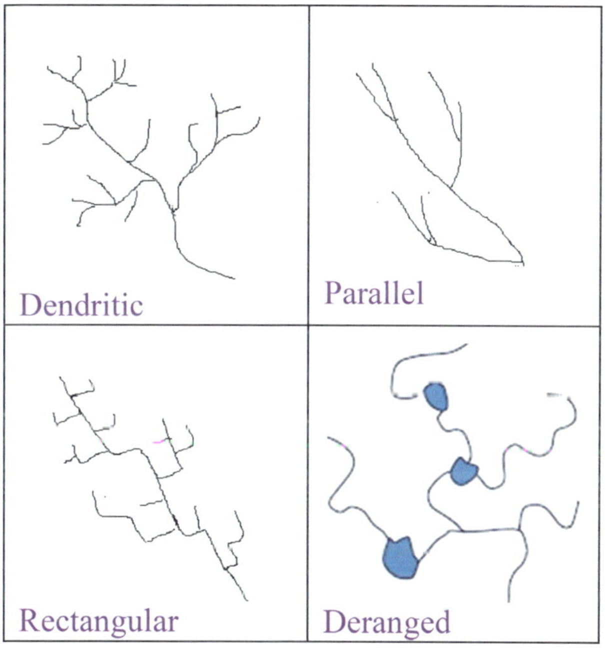

Figure 2.
Drainage patterns commonly encountered at mining sites.

according to their size] has been used for planning and management [9]. This term "watershed" is taken synonymously with catchment or drainage basin, an area of land which drains to a common outlet, and is said to be related with water only. But watershed and its management are not only related with water; it essentially relates to resource conservation which means proper land use, protecting land against all forms of deterioration, building and maintaining soil fertility, conserving water, proper management of water for drainage, sediment reduction, and increasing productivity from all land uses. According to the multilevel planning policy at national, state, district, and lower area levels, natural resource data management, which includes groundwater as well, is done on watershed basis considering each watershed as a constituent unit for planning. Thus, to facilitate area-specific microlevel planning for management of water resources (groundwater/surface water), it is convenient to apply "integrated strategy" on watershed basis. This integrated approach has close relation with watershed and management related therewith because it defines the optimum conservation of water with due regard to other resources. Watershed approach provides the mineral production which is resource-centered and environmentally friendly and helps in promoting sustainable development and pollution abatement.

2.3 Control and treatment

Several techniques of "control and treatment" are available to manage groundwater, for example, zero-level discharge. As a basic rule of thumb, if one has to control and treat water in a mining area, the approach should be to keep pollution contained in the mine itself. Their control beyond the mine boundaries is neither economical nor manageable. Depending on the problem encountered, groundwater infiltration or discharge should be handled and aquifer contamination be avoided, for example, oxidation and leaching of mine drainage produce high iron and sulfate concentrations and low pH in groundwater.

For the sustainable development of mining areas, the main source of pollution should be traced, and by applying chemical and bacteriological methods of treatment, water pollution shall be dealt or treatment methodology applied. The cost of treatment and risk involved must be checked for viability of adopted measures deployed to control pollution. To control mine water discharge and treat it for pollution abatement, *Intelligent Mine Water Management* (*IMwM*), a solution for mine water management, is extremely scientific in approach (**Figure 3**), which can be enforced by the mining industry, regulators, and stakeholders world over.

As depicted in **Figure 3**, *IMwM* when expanded fully (see points below) will explain past, present, and future of control and treatment thereby helpful in maintaining an acceptable standard of living—now and in the future for benefits to the mining industry. All technical interest related to water in mines, including the burning and current topics, are covered under the following:

- Mine water hydrogeology
- Mine water geochemistry
- Qualitative analysis of mine water
- Quantitative estimation of mine water
- Water-related mine design (tailing pond, etc.) and dewatering planning

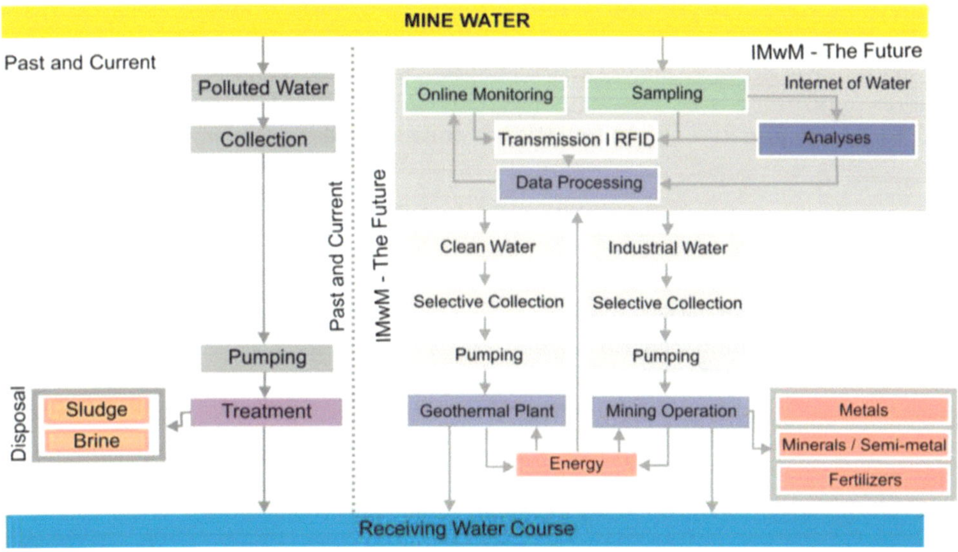

Figure 3.
Intelligent mine water management (IMwM) (courtesy: Christian Wolkersdorfer, IMWA).

- Mine water utilization and end use of mine water (extracting values from mine water)
- Mine water pollution (from tailing, dumps) and mine water discharge
    - Mine water monitoring and treatment
    - Microbiology of mine waters and bio-leaching
    - Stable isotopes in mine waters (tracer test, etc.)
- Mine water management (approach, strategies, and social conflicts)
- Mine closure, remediation, and follow-up care
    - Mine water limnology/pit lakes
- Geotechnical issues related to mine water (destabilization of slope/slope failure)
- Mine water modeling (three-dimensional or two-dimensional)
- Process simulation tools related to mine water
- Water policy issues in mining
- Mine water regulation
- Mine water and climate change

Best mining practices (BMP) to curb and contain mine water pollution, groundwater lowering, radius/area of influence, groundwater recharge, induced infiltration, cone of depression, water table lowering, mine drainage, consequences of dewatering and management, etc. are all covered in it.

Recycling concept rationally articulated for comprehensive short-term as well as long-term planning is very useful for water control, treatment, and management provided their effective implementation is done in field.

3. Impact of mining on groundwater (mining and its consequences)

Mining of minerals often leads to various environmental impacts [10] including water [11, 12]. The analysis of impact(s) can be done by comparing present scenario with past or pre-mining scenario [13] and evaluated as either positive or negative or the combination of both. These are analyzed with respect to the core zone, 5 km radius area (or alternatively consisting of the active mining area alone), and buffer zone, consisting of 10 km radius area. The likely impacts of open-pit mining could be in terms of:

a. Drawdown, that is, lowering of water table

b. Water quality deterioration, that is, water pollution

3.1 Groundwater lowering due to water table interception

When water is discharged from the pit mine which has intercepted water table, firstly the "collected water" is discharged, and then water from phreatic surface (water table) is sucked, and a "cone of depression" is formed with its axis at the lowest point at the sump bottom having lowest RL. If discharging is done for more time period, this cone of depression continues to enlarge, and pronounced effect is noticed. In technical terminology this is what is referred as "drawdown." **Figure 4** explains this drawdown principle in general for a discharge through pit or dug well as applies in hydraulics. If more than one point of water discharge or drawdown exists in a pit mine and kept overlapped, the lowering of water level takes place rapidly, and quarry bottom can be dried with faster speed (**Figure 5**).

In respect of drawdown, two different kinds of situations come across in an open-pit mine: firstly, when the mine is working above the water table and, secondly, when the mine is working below the water table. Water (or drawdown) does not pose any problem in the former case, whereas in the latter case, lowering of water table may be the impact of mining. As a general principle, drawdown is usually in excess of 65% of

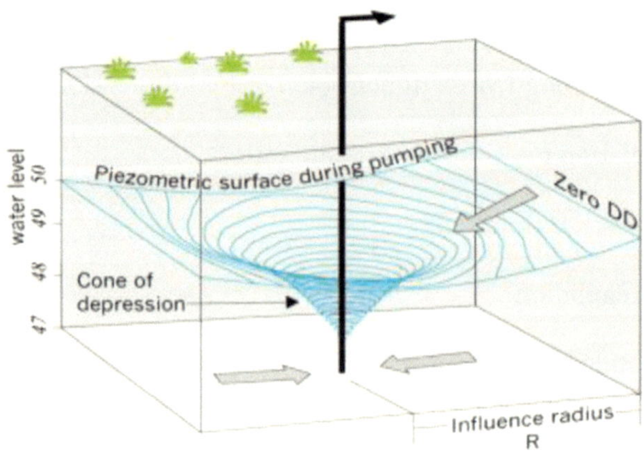

Figure 4.
Cone of depression and radius of influence (courtsey: Dr. Yohannes Yihdego, Australia).

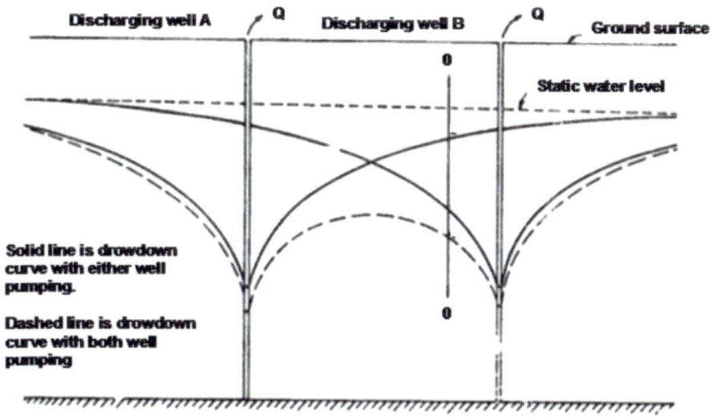

Figure 5.
Interference between discharging wells (courtsey: D.K. Todd).

unconfined aquifer thickness [14]. Such drawdown varies from rock type to rock type. Therefore, this statement cannot be taken as a thumb rule.

While analyzing the impact of mining on nearby villages, that is, adjacent to pit, the water-level records or fluctuations (in open dug well/borewell or piezometer) in pre-monsoon and post-monsoon season are taken into account. Because India has monsoonal climate and maximum rainfall occurs during June to September months, pre and post monsoon philosophy is considered best. On the basis of field observations, that is, rock, formations, and aquifer conditions, the impact is assessed of that study area, for example, GW in hard rocks will be present in the fractures/cracks/ and fissures in small quantity while compact soft sandstone rocks contain significant groundwater quantity in rock pores and interstices. The continuity of cracks in aquifer determines the water availability even though stratum has impervious characteristics. Therefore, in such situations drawdown by pumping will be observed as local impact only. Another impact of mining that could be natural also is defined in terms of "radius of influence."

It is often asked how to estimate or quantify the impact of mining on groundwater regime? This question can be scientifically and effectively answered by estimating the influence radius or radius of influence (Ro/Re). The importance of Ro/Re with respect to a mine is that it demarcates a visually assessable picture of impact in terms of a measurable distance and should be kept constant/or minimum as far as possible.

Radius of influence (Ro) in technical terminology is the impact area, spread around the mine due to groundwater extraction or use. It is calculated using Eqs. (1) and (2) given in the below figure.

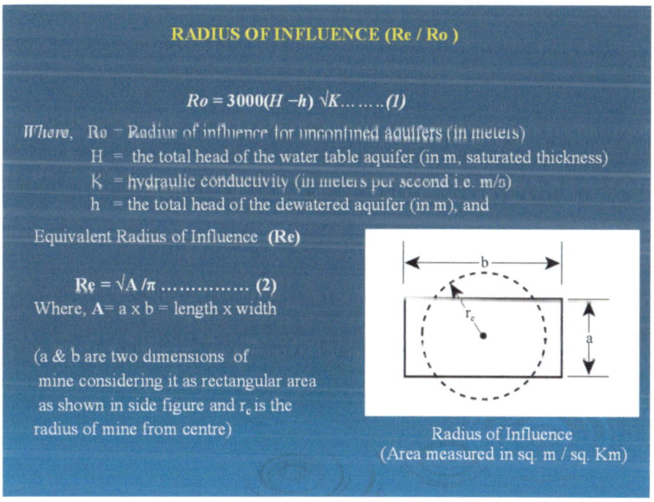

Ro is directly proportional to "draft magnitude" and "average rainfall" that occurs in an area. Here, the GW extraction is limited to mines only and as an industrial unit which otherwise could be for irrigational, agricultural, or domestic purpose also. For a "single pit" in an open mine, an equivalent radius of influence (Re) is calculated, whereas "Ro" is determined for multiple/concentric pits.

The operative staff, for all practical purposes, can judge the cone of depression, drawdown conditions, and radius of influence in the mine based on their field experience.

3.2 Water quality implications

Besides groundwater lowering, water quality implications (in the form of pollution) are a major impact issue of mining on environment globally. The pollutants (or traces of heavy metals) are released into the groundwater by geogenic sources through weathering of the geologic formations [15] and anthropogenic sources. Contamination in groundwater because of anthropogenic sources, for example, agricultural fields and use of fertilizer/pesticides, sewages and solid wastes, return flow due to irrigation, etc., is most often noticed and is far-far larger than the water-level lowering impact mentioned above. The water quality implications and environmental impacts are described/covered in appropriate section of this chapter in a scattered manner. It is so because number of cross-connecting factors of land and water has to be looked into for quality evaluation (Sections 2 and 5.2).

4. Groundwater and planning for mining below ground level

When mine becomes deep or excessive watery conditions are encountered in underground mines or when mine is located in the vicinity of a major water body and intensive seepage through strata (more than normal) occurs, then scientific mining and planning for groundwater management becomes essential. At varying locations, different mining and differing groundwater conditions are observed, for example, when mine is located adjacent to sea/in coastal areas, when aquifer encountered is confined and water table is under pressure, etc. In all these situations, mine planning for mineral extraction below the water table has to be carried out differently taking into account the water hydrology. World over, the depth denomination differs from country to country for an open surface mine, operating in pit form (**Box 1**). But in general and in practical sense, all mines below water table are likely to encounter water or watery condition whether it is an open-pit mine or an underground mine.

Planning for mining below ground level has to consider the effect of deepening of pit. Therefore, an interdisciplinary approach intermingling both planning and engineering aspects is needed. Considering the constraints posed by the dynamics

Box 1. When mine is deep?

Deep mine or deep mining is simply mining underground, in which the miner and/or machinery work beneath a cover of soil or rock. There is no fixed norm for mine to become "deep" or "shallow." As a rule of thumb, exploitation of fuel minerals (coal/lignite/brown coal) at depths exceeding 300 m depth can be considered as deep mine, whereas for metallic or nonmetallic mineral deposits of modest mineral value, this norm may be taken as 350 m approximately.

When open-pit mine is deepened beyond a certain depth, "economic stripping ratio" comes into picture and the underground mining originates in which gaining access to the mineral deposit is by means of vertical shafts, inclined shafts, drift mining, or by other means. The value of mineral exploited, that is, cost of mineral production from mine (ROM cost), govern its excavation depth. For a higher value mineral and lower value mineral, such norms are staggering differently.

of groundwater, that is, spatial variability, hydrogeological data and its availability, socio-economic conditions, demographic profile of the area, etc., its quantitative estimation is done. In Indian condition, Groundwater Estimation Committee (GEC-1997) methodology seems practical for calculating water quantity. On this basis, planning of mining below ground level and water management through engineering approach yields desired output. To plan a mine for industrial purpose, obtaining groundwater abstraction permission is necessary. Such statutory compliance, particularly groundwater permission in mining, makes the water management easier [16]. In India and until now, it was mandatory for all new industries to apply for groundwater extraction clearance, but now it is mandatory to obtain these clearance for old as well as new industry (http://times ofindia.indiatimes.com/articleshow/49832855.cms?utm_source=contentofinterest&utm_medium=text&utm_campaign=cppst). This has initiated the need and emphasized for estimation of groundwater quantity and its management.

To do the planning as per the approved mining plan, excavation depth (RL/MRL) and the lowest MRL up to which mining will reach in the future have to be designed scientifically. Depth-wise RL, pit dimensions, and water quantity (Q) are then needed for assessment. It may be noted that the excavated area dimensions keep on changing as per the *ultimate pit plan. A*s per the dug-out area, the water availability in the mine area varies during different periods of a year. Accordingly, water quantity (Q) is first estimated for that particular mining pit. Related to Q or water quantity, three areas are important, namely, "mine lease area," "catchment area," and "pit area" (excavated area/water-filled area).

Geohydrological evaluation of the mine area is extremely helpful for the groundwater assessment and futuristic planning of the mine area. In addition to the GW and SW, seepage water is also accounted for in mine's planning. Seepage water appears through mine walls in open pits, and field observations for seepage flow are generally recorded during post-monsoon season. To get the total water quantity of mine pit, it is simply added to the SW and GW quantity.

By groundwater modeling and simulation methodology, groundwater-level decline (maps, etc.) and the groundwater quantity can be estimated [17]. To understand the groundwater resource position in a mining area, water table depth below ground level and aquifer types are extremely important. If these are known and utilized correctly, the planning for mining will be easier. A general trend indicating rate of groundwater discharge/rate of outflow with time is illustrated below (**Figure 6**). The help of graph can be taken to know the availability of water during different months in a year, which varies from 200% (100% for surface water and 100% for groundwater) to as low as 55%.

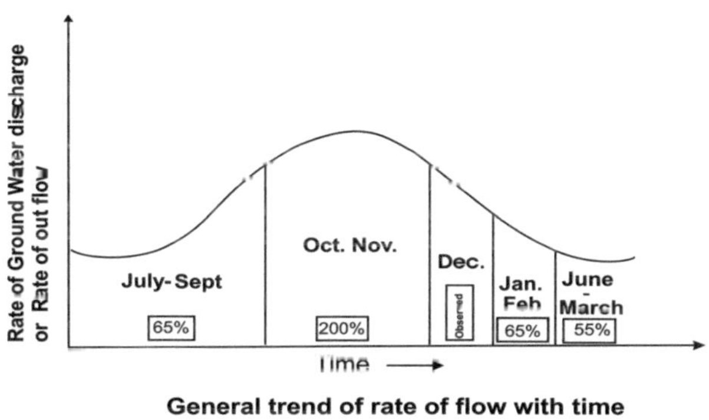

Figure 6.
Rate of groundwater discharge (or rate of outflow) with time

As a part of mine planning, operation, and execution, following methodology is helpful for mining *below ground level*:

1. Based on topo sheet of the area, a *drainage map* is prepared. These days using GIS and satellite imageries also, such maps are prepared.

2. Observe the flow pattern of surface water in and around the mine area.

3. Determine the groundwater flow and its direction using hydrogeological map of the area. Such maps are also available with state/federal groundwater authorities.

4. Calculate total water quantity which includes SW + GW+ seepage water.

5. Find out the "area of influence" in measurable parameters, and assess the real field conditions.

6. Execute planning of mine, keeping in mind the drainage, flow direction, water quantities, influenced area, and ground elevations of various nodal points of mine lease area.

7. Make "dewatering plans and scheme" in totality and not patch-wise. Sump design with "desilting arrangements" and suitable "pumps and pumping system" are a part of dewatering plan and scheme [18]. Their design should be based on engineering considerations and technical intricacies.

8. Decide the network of drains and drainages, its location, elevation, etc. for proper water outflow of water based on drainage pattern of the area. Make use of surveying for finalization.

9. Keep check on:

    a. Runoff inside the pit

    b. Slope erosion and control (including stabilization by natural vegetation, etc.)

    c. Sediment/silt load accumulation in sump/sedimentation pond, etc.

    d. Water quality and its deterioration at mine level

    e. Periodical maintenance/observation

10. Prepare a "master plan" and implement it in practice.

5. Quantitative estimation and qualitative analysis of groundwater

Mine water in the mining areas comes across two broad issues, namely, water quantity and water quality. In most of the mines and in different parts of the world, both quality and quantity of groundwater resources are required for management, for statutory compliance, and for planned extraction of minerals from the mine [19–23].

The groundwater resources have both static and dynamic dimension. But essentially it is a dynamic resource which is replenishable (annually or periodically) through precipitation. It is static in "saturated zone" and dynamic in the upper

DOI: http://dx.doi.org/10.5772/intechopen.85309

unsaturated zone (upper part of the water table) where water-level fluctuation is recorded. Near accurate estimation of groundwater resources is possible by adopting a set of the steps and formula framed for the purpose. A brief about Indian methodology for groundwater estimation is given in this chapter below for reader's knowledge and understanding. This may be noted that from country to country, such estimation procedure or methodology may differ.

To estimate groundwater extraction in an open mining pit, two broader approaches are possible. First is "planned depletion approach" (sustainable yield method), and second is "safe yield approach." "Safe yield approach of assessment" is based on groundwater recharge that takes place in an area or region, and recharge is calculated using *water balance method*, discrete numerical modeling, or tracer technique. In the "sustainable yield method," assessment can be made using "discrete numerical modeling" only. In India, later one safe yield approach is adopted and found more appropriate for groundwater estimation. Based on this approach, groundwater estimation methodology (GEC)-1997 has been formulated by Central Groundwater Board (CGWB), Govt. of India, and the same is applied in Indian mining sector for groundwater assessment.

5.1 Estimation of groundwater quantity (Q)

Groundwater Estimation Committee methodology, abbreviated as *GEC '97* methodology, is an interactive methodology designed by the expert committee [24]. India has adopted it for estimation nationally, and since then mining water quantity is also estimated by this methodology. For groundwater estimation in India, methodologies for alluvium/soft-rocks and for hard-rock areas both have been formulated by the expert committee. This is significant to note that nearly 80% of the mine areas lie in hard-rock terrain.

India with its vast areal extent, long coastline, and large deltaic tracts forming a linear strip around peninsula is characterized by diversified geological, climatological, and topographic setup. Discontinuous aquifers of varying yield potentials occupy 2/3 area of the country, and as said above most of the mine area lies in hard-rock terrain. Thus, GEC '97 methodology and its norms for hard-rock areas [24] remain applicable for evaluation and assessment of groundwater. By understanding the behavior and characteristics of rocks, the water quantity as well quality in the mining area can be estimated. Steps and formulas of GEC '97 methodology and the calculation for open-pit mine (surface mine only) are shown below.

(A) Groundwater calculation

GW quantity available is that quantity which is likely to be experienced in the form of pit water either as punctured water table (groundwater) or in the form of seepage water from the footwall (FW)/hang wall (HW) sides of mine pit walls (see point C of this section below).

- Groundwater Quantity (W1)

(for mine lease area and maximum rainfall/maximum water-level fluctuation occurred for worst-case scenario)

Method 1: infiltration method

Maximum feasible groundwater quantity

$$A\ (m^3) = \text{lease area/pit area} \times \text{rainfall (max.)} \times \text{RIF}$$

(refer Table 1 for rainfall infiltration factor (RIF) values)

Method 2: specific yield method
Maximum feasible groundwater quantity

$$B\left(m^3\right) = \text{lease area} \times \text{max.fluctuation} \times \text{specific yield}$$

Average groundwater quantity within lease hold area in a year

$$C = (A + B)/2 \left(\text{in } m^3/\text{TCM}/\text{MCM}\right)$$

Considering, 365 days in a year, quantity in a day can be worked out

$$\text{Thus, available groundwater quantity in } m^3/\text{day} = W_1/365$$

Note: *Groundwater, as base flow, is present in the mine area during whole year, and seepage is governed by the geological and topographical features of the area. Thus, groundwater availability can be taken as 365 days in a year.*

- Groundwater development/groundwater utilization for mine area

Groundwater development can be assessed and estimated by the established procedure of GEC '97. An assessment about the stage of groundwater development is helpful in knowing the overall groundwater scenario of the study area.

The stage of groundwater development in a given sub-unit is defined as the current annual gross groundwater draft for all uses (C) in that sub-unit expressed as a percentage of the net annual groundwater availability (B) in that sub-unit (GEC '97). Thus, if stage of groundwater development is "A," this can be calculated as follows:

$$A = \left(\text{gross availability}/\text{net availability}\right) \times 100 = (C/B) \times 100\%$$

Similarly, for a mine area GW utilization = output/input (in percentage) = total discharge through mine/net groundwater availability

Category	Stage of GW development	Water table trend/level
Safe	≤70%	No water table falling trend
Semi-critical	>70% ≤ 90%	Falling water table trend
Critical	90–100%	Falling water table trend
Overexploited	>100%	Falling water table trend

The sub-unit for the purpose of assessment can be a lease area of mine or a command/non-command area. Having known the GW development/utilization in the mining area, the same can be compared with the standard regional norms. Based on this, the very purpose of evaluation and assessment of groundwater analysis can be categorized as "safe" or "critical."

According to the availability, the current stage of development, and water table fluctuation trend, its allocation for various uses in future, that includes domestic and industrial uses, can be made.

- Groundwater recharge or total annual replenishable recharge (TARR) (unit—m^3/TCM/MCM)

This is the maximum feasible recharge per annum (Rc or Rc'), and usually referred as total annual replenishable recharge (TARR) is calculated by two methods as per the formula given below.

Method 1: rainfall infiltration method

$$\text{Rc} = \text{catchment area} \times \text{rainfall (average)} \times \text{rainfall infiltration factor}^{*}$$
$$\text{or Rc in million m}^3 \text{ (MCM)}$$

Method 2: specific yield method

$$\text{Rc}' = \text{catchment area} \times \text{water table fluctuation (average) x specific yield}$$
$$\text{or Rc}' \text{ in million m}^3 \text{ (MCM)}$$

Note:

i. *Here * = rainfall infiltration factor (RIF) = values as per GEC '97 (**Table 1**).*

ii. *For TARR calculation catchment area or alternatively the active mining area can be taken.*

iii. *Normalization of rainfall recharge: the water table fluctuation in an aquifer corresponds to the rainfall of the year of observation. The rainfall recharge estimated should be corrected to the long-term normal rainfall for the area. For calculating the annual recharge during monsoon, the formula indicated below is adopted as per GEC '97 methodology.*

(A) For alluvial terrain of India				
S.no.	**Geographical location/formations**	**RIF as a fraction**		
		Recommended value	**Minimum value**	**Maximum value**
1.	Indo-Gangetic plains and inland areas	0.22	0.20	0.25
2.	East coast	0.16	0.14	0.18
3.	West coast	0.10	0.08	0.12

(B) For hard-rock terrain of India				
S.no.	**Rock types**	**RIF as a fraction**		
		Recommended value	**Minimum value**	**Maximum value**
1.	Weathered granite, gneiss, and schist with low clay content	0.11	0.10	0.12
2.	Weathered granite, gneiss, and schist with significant clay content	0.08	0.05	0.09
3.	Granulite facies like charnockite, etc.	0.05	0.04	0.06
4.	Vesicular and jointed basalt	0.13	0.12	0.14
5.	Weathered basalt	0.07	0.06	0.08
6.	Laterite	0.07	0.06	0.08
7.	Semi-consolidated sandstone	0.12	0.10	0.14
8.	Consolidated sandstone, quartzite, limestone (except cavernous limestone)	0.06	0.05	0.07
9	Phyllites, shale	0.04	0.03	0.05
10.	Massive poorly fractured rock	0.01	0.01	0.03

Table 1.
Rainfall infiltration factor (RIF) as per GEC '97 and terrain conditions.

$$\text{Monsoon recharge} = \text{area}\ (\text{km}^2) \times \text{water} - \text{level fluctuation}\ (\text{m}) \times \text{specific yield.}$$

Groundwater recharge may also take place through other point/line sources namely tanks, ponds and river/nala. Thus, recharge through different sources includes:

a. Recharge through irrigation

b. Recharge through stagnant water bodies namely ponds and tanks etc.

c. Recharge through water-filled small-sized pits or spread of water

d. Recharge through return flow

This can be estimated for the catchment area and command/non-command area as the case may be using GEC '97 methodology. Its descriptive details can be referred from [25]. With increasing focus on sustainable development of groundwater resources, augmentation of water conservation structures, with the aim of increasing groundwater recharge, can be implemented in the field. The water conservation structures include percolation tank, check dam, nalla-bund, etc. Recharge through such planned/proposed recharge structure can then be calculated by knowing average water spread area, seepage factor, and water containment days.

- Draft calculation/estimation (unit—m^3/TCM/MCM per year)

"Draft" means consumption. In mining case study, which is an industrial setup, three types of drafts are considered prominent in estimation/calculation, namely, "domestic draft," "draft through mine discharge," that is, pumped out water quantity from pit, and "industrial water draft" for mineral processing (consumption). To estimate domestic draft, total population of the study area and sources of groundwater abstraction must be known. For such calculation all villages and human settlements in the core zone (CZ) and buffer zone (BZ) area are considered which covers 10 km radius area around the pit center. Thus, domestic draft/year (considering groundwater as the only sources).

$$= \text{Population/no of persons (total)} \times \text{water consumption per head}$$
$$(\text{liter/person/day}) \times \text{days in a year}$$

(B) Surface water calculation

In general, surface water (SW) quantity, that is, W_2, is calculated on per day basis because surface water quantity differs from season to season. This quantity is dependent on rainfall/precipitation (during wet season of monsoon). For estimation purpose, maximum rainfall occurred (i.e., worst-case scenario) or average rainfall for 10 years or more can be considered. Separate estimation should be done/shown for peak rainfall period indicating number of days and either the lease area or catchment area as the case may be is considered for calculation.

Surface water quantity (W2)

$$\text{Surface water quantity/day} = W_2 = \{X–(Y1 + Y2)\}/\text{no.of rainy days}$$

where $X = [(M1 + M2) \times rainfall]/2$

> *where M1 = lease area/catchment area; M2 = water filled area; Y1 = evaporation losses (30% of the rainfall); Y2 = infiltration losses (10% of the rainfall).*
>
> *Y1 + Y2 are the "water losses," which are taken into account for the estimation/calculation. Nearly 40% of the rainfall goes as waste in the form of "total runoff" for the hard-rock areas.*

In India, where monsoonal climatic condition exists, the maximum surface water quantity in a mining pit will be available for a period of 92 days (3 months approximately) in a year, that is, during monsoon and post-monsoon period of July to September end only. In summer season, quantity of water present in pit as well as in lease area will be minimum and always less than the quantity during monsoon period (**Table 2**).

(C) Seepage water calculation

Normally, seepage water in mine pits occurs as a result of interconnection of pit wall with water body located either in vicinity or at a distance. Capillary action with aquifer also leads to the seepage on pit walls even at upper elevation. If less seepage is observed, the same can be ignored, and seepage water quantity can be taken as "nil." For more seepages, the calculations are based on the general principle of water outflow from the seeped surface area in a recorded time. It is simply added to the SW and GW quantity to obtain total water quantity. Thus, seepage water quantity *(Qseepw)* of mine pit is equal to flow rate in a given time of that surface area from where seepage is occurring.

(D) Water balance

When GW, SW, and seepage water quantity is known, the water balance of the assessment area is calculated as follows:

S.no.	Period/ month	Total number of days	Availability of water quantity		Remarks
			Surface water	Ground water	
1.	January to February	59 days	Less than moderate	Present as base flow	1. Surface water, groundwater, seepage water, and water through recharge constitute the total water quantity present in the area 2. Maximum means peak season quantity; moderate means 75% of peak quantity; less than moderate means 30% less than peak quantity. Minimum means 50% less than peak quantity
2.	March to June	122 days	Minimum	Present as base flow	
3.	July to September	92 days	Maximum	Maximum	
4.	October to November	61 days	Less than maximum (but ample)	Maximum	
5.	December	31 days	Moderate	Present as base flow	

Important notes:

1. *Since, surface water quantity varies widely over the time period of one complete year, total yearly calculation should not be indicated/shown.*
2. *Groundwater and surface water quantity and availability shown above are applicable for Indian mines/geo-mining condition and elsewhere the situation may change according to the pattern and period of precipitation.*
3. *Groundwater availability (W_1) and surface water quantity (W_2) in m^3/day should be shown separately indicating the area and period considered. Summation of groundwater quantity and surface water quantity of the study area should not be done because groundwater and surface water quantity and availability fluctuate throughout the year.*

Table 2.
Quantity-wise surface water availability over different periods in a year

$$\text{Water balance} = \text{input} - \text{output} = \text{net GW availability} - \text{discharge through mine}$$
$$= (+) \text{ or } (-) \text{ can be expressed in TCM or MCM/year}$$

Case records: To know the field scenario, four open-pit mines of India are studied, and they are (i) Partipura limestone mine, Banswada district, Rajasthan [26]; (ii) Rajhara iron ore mine, Balod district, Chhattisgarh [27]; (iii) Malanjkhand copper ore mine, Balaghat district, Madhya Pradesh [28]; and (iv) Lanjiberna limestone and dolomite mine, Sundergarh district, Odisha [29, 30]. In all these mines, different minerals are excavated, and varying geo-mining conditions exist. At *Partipura limestone mine*, mining conditions are that of a normal open-pit mine, whereas twin mining, that is, surface mining and underground mining, both exist in close vicinity at *Malanjkhand copper ore mine* and *Rajhara mechanized mine* of Chhattisgarh state, the excavated iron ore is very hard, and ore reserves are getting exhausted, that is, mine has reached at its last stage of life. "Lanjiberna mine" of Odisha is a typical surface mine in which three pairs of pits (i.e., total six dug-out areas) are excavated for obtaining limestone and dolomite, which is filled with water. All these working pits have different depths (**Table 3**), and water table is intercepted as a result of mining. At all these mines, comprehensive geohydrological studies had been carried out, and groundwater quantity using GEC '97 is assessed (**Table 3**). It is observed that all the four mines are having hard-rock formations with unconfined aquifers (GW occurs under water table condition). Average rainfall and maximum rainfall (for the worst-case scenario) in all these mines differ widely, and average water-level fluctuations (WLF) are less than 10 m below ground level. In different seasons the water quantities fluctuate widely for which number of reasons and factors are responsible. When water quantity Q is checked (verification by ground truth), it was found correct by the concerned statutory agencies with ±10–20% variations from actual. Based on this, necessary permission for continuance of mining operation was granted for these mines.

5.1.1 How to estimate Q for an underground mine ($Q_{u/g}$)

Having discussed the groundwater quantity for an open-pit mine, an obvious question arises. Whether groundwater calculation for underground mine ($Q_{u/g}$) is also estimated in the same way? Its answer is no. The approach for estimating groundwater quantity with respect to an underground mine is sharply different. Q for an u/g mine is full of uncertainties and based on the actual field conditions encountered. Such field conditions are many, either created or naturally encountered, for example, extent of underground mine development affects the creation of void's underground, this in turn has a close connection with groundwater movement in encountered aquifers.

Secondly, depth of underground workings from surface has linkages with groundwater recharge occurring in that particular area, which in turn is related with local rainfall. Obviously, rock types, its porosity, and hydrological characteristics have key role in groundwater movement. Similarly, geological features such as faults, folds, unconformities, lineaments, etc. reflect their own dominance in groundwater quantity as well as movement. Thus, both rock type (different formations) and geology, for either open-pit mine or an underground mine, have tremendous importance. Its detailed study and engineering judgment can help one to estimate the groundwater quantity approximately, if not exactly. Thus, approach for estimating $Q_{u/g}$ must incorporate study of borehole litho-logs of the mine/area and other related parameters, namely, rainfall, recharge, aquifer and its characteristics, extent of underground mine development, and working depth. Based on

(i) Partipura limestone mine [26]	(ii) Rajhara iron ore mine [27]

Annual GW quantity at 70 m pit depth = 0. 1616 MCM	Annual GW quantity at 250 m pit depth = 1.713 MCM
(iii) Malanjkhand copper ore mine [28]	(iv) Lanjiberna limestone mine [29, 30]

Annual GW quantity at 240 m pit depth = 4.735 MCM Note: MCM = Million cubic mèter; GW = Groundwater	Annual GW quantity for pit depth from 39 m to 56 m = 0.268 MCM to 1.4 MCM [pit depths: pit 1 and pit 3 = 41 m; pit 2 and pit 6 = 56 m and pit 4 and pit 5 = 39 m]

Table 3.
Estimated GW quantity for different case studies using GEC '97

groundwater movement principles (Darcy's law), runoff and recharge relationship of surface water and general estimation formulas as applied in GEC '97 methodology $Q_{u/g}$ for quantitative can be estimated. Further, this may be noted that the "underground mine water quantity" is proportionally related with the actual excavation area exposed in underground workings (size and area of panel/stope) as well as surface area above the extraction/depillaring panel.

Here, it is important to reaffirm that in the paragraph above, author has clearly showed how the groundwater quantity can be calculated and how it is related with several factors. This water quantity calculation is helpful at the planning stage and operational stage of the project for "dewatering planning and related aspects." One can also know the *water availability* and how to use it whether within mine pit or outside. Similar excavations being operated below ground, for example, caverns, tunnels, etc., are the other beneficiaries for such knowledge. Since the estimated quantity(ies) are based on aquifer parameters and scientifically proven, it is true and near actual. Its immense benefits can be encashed, in terms of cost savings and cost overrun of project(s).

5.2 Qualitative analysis of groundwater in mines

The qualitative assessment of groundwater samples (or surface water samples) from the mining area and surrounding areas is required to infer water quality (WQ) and thereby knowing its suitability for various uses. The mine water vary greatly in terms of concentrations of various chemical constituents, as water quality is likely to be affected with mine site parameters which are specific in nature.

Various studies on interrelationship between water quality, geology, and mining activities have been carried out in Indian mines [11, 15, 21, 31–33]. Similar studies and attempts are in vogue with reference to the different mines around the world, and their list is exhaustive.

Water quality assessment can be done either by field method(s) or by laboratory method(s). Their related aspects, that is, quality parameters and its selection for analysis, characterization, field sampling, water storage before and after lab analysis, periodical monitoring of quality for drawing inference, etc., require in-depth description parameter-wise [34]. For this, standard operating procedure (SOP) can be applied [35], and available literature on the specific subject can be referred for the details. Their elaborate description (field and laboratory method) has not been described in this chapter because ample of literature is available on the water quality and its assessment, even some of it is described by other authors in this book itself.

As regards water quality in mines, the following comes into the reader's mind—(i) the water quality of surface channels flowing in the mining area; (ii) the mine pit water quality; (iii) dump/spoil bank water quality; and (iv) tailing ponds/impoundments water quality. Depending on the type of mineral excavated, the quality issues are to be recognized and assessed, for example, acid mine drainage problems are a severe water quality issue in the case of coal mines. The elemental analysis (pH, TDS, total hardness, etc.) is needed for limestone and dolomite mine, whereas lead-zinc, copper, iron ore, and bauxite mines (mines of metallic minerals/ore) require attention toward heavy metal constituent's analysis.

Inseparable surface water and groundwater and its pollution can be assessed or evaluated qualitatively. Some important water quality parameters and major possible water contaminants and pollution indicators for mining and allied industry are shown in **Tables 4** and 5. Having determined the value of each parameters, may be either low/high or within permissible limit/outside permissible limit, the scientific explanation of pollution status can be given. For each of the studied case, the pollution parameters that accounts are different and as according to the water usages. It is also recognized that the mine water quality, which is present in the mine or in the surrounding areas around the mine sites; in shallow aquifers and deep aquifers of mine sites, though not comparable with one another but governed by the same scientific principles/groundwater chemistry. Chosen WQ parameters are hence critical for valuation.

By knowing the water quality, one can easily trace back the source(s) of pollution, and management measures can be taken accordingly. Further, it is helpful and suggestive to know the background history also for proper assessment of WQ. Advances in instrumentation, modern computational technology, and improved management techniques are able to reduce many negative impacts arising out of water quality pollution.

It is found that the pH of the mine water fluctuates both ways from the normal range of 7 and the total hardness (TDS) parameter also varies considerably depending on the prevailing hydrological regime and the variation in lithology. These parameters mainly decide the mine water suitability for domestic, irrigation, and other miscellaneous uses. The anion and the cation chemistry (dominated by HCO_3^-/SO_4^- concentration and Ca^{2+} and Na^{2+} ions, respectively) and hydrochemical facies (Mg-Ca-HCo_3 and Mg-Ca-HCO_3-Cl, etc.) knowledge can put forward the water chemistry mechanism for its various uses [36]. By knowing parametric values of various chemical parameters, sodium absorption ratio and residual sodium carbonate and acidity/salinity of mine water can be determined or assessed.

Physical	Chemical	Biological	Radiological
Color	pH	Virus	Uranium
Odor	Acidity/alkalinity	Bacteria (*Coliform*)	
Temperature	Hardness	Algae	
Turbidity	Ammonia (free)	Other nuisance organisms	
Foam and froth	Nitrates		
	BOD/COD	Human-related inorganic constituents	
	Calcium	Arsenic	Lead
	Magnesium	Asbestos	Zinc
	Chlorides	Barium	Nickel
	Sulfates	Cadmium	Nitrate
	Phosphates	Chromium	Selenium
	Sodium	Cyanide	Silver
	Potassium	Fluoride	Sodium
	Redox potential	Iron	Mercury
	Conductivity	Hardness	

Table 4.
Some important water quality parameters

TDS	Sulfates (SO_4^{2-})	Fluoride
COD	Arsenic	Phosphates ($PO_4^{?-}$)
BOD	Bicarbonates	Zinc
Carbon (organically linked)	Iron	Copper
Hydrogen (organically linked)	Manganese	Lead
Nitrogen	Sodium	Mercury
Detergents	Potassium	Temperature
Oxygen	Calcium	pH
Nitrates (NO_3^-)	Magnesium	Conductivity
Nitrites (NO_2^-)	Total hardness	Redox potential
Ammonia (NH_4^+)	Chlorides	
H_2S (in dissolved form)		

Table 5.
Main possible groundwater contaminants and pollution indicators in mining and allied industry

Thus, in brief, it is learnt that assessment of water quantity and quality is a pre-requisite for planning and development of mine. Mining of minerals at shallow depth can be done without adversely affecting the groundwater; however, when mine/mining goes deep, that is, below water table, the need to check its quantity, availability, and scientific management arises. Both quality and quantity assessment parameters are since field-based; a minor departure in filed values is possible. Nearly (±) 20% departure from actual scenario is generally observed and admissible. By overcoming field measurement difficulties and adopting standard operating procedure (SOP), near accurate evaluation of groundwater can be done.

6. Policy framework for mines

For the mining industry as a whole, clear and transparent "corporate policy" provides a direction for the implementation of plan and programs in respect of water. This effectively controls the cost component as well and expresses the desire of the organization to achieve the aims fixed toward the improvement of water management. If any organization policy recommends for better water utilization and sound water management, then it is also essential that companies must have ingenuity for its effective implementation. Commonly, policies do exist in developing countries, but the desire for their implementation is often lacking. This is particularly the case in small- and medium-sized companies in "unorganized sectors" having lacks of financial resources. One of the difficulties, mining companies or the mine management focuses in regard to policy formulation are also the lack of proper equipment, machinery, expert knowledge, or financial resources for executing the policy. Adequate funds are essential for the implementation of plans and ideas. Furthermore, the organization policy and the national water policy should be in tune with each other [37].

It should be emphasized here that water policy usually focuses on water in general and not in particular on mine water. This becomes especially complicated when the water policies and the mineral resources policies are managed by different departments or regulatory bodies in developed and developing countries. Deficiencies in certain aspects of groundwater-related policies particularly on the management aspects and its core issues can be addressed as well as enforced by the industry, regulators, and stakeholders through policy perspective.

7. Extracting value from mine water

Before I discuss extraction of value from the mine water, let me clarify that it is beyond doubt that water is everybody's concern; it is apparent that water in general or water from mine (s) should not be wasted and has to be properly utilized as well as conserved. Different aspects of water utilization and conservation are commonly dealt in books, but how to earn or extract value from water is dealt very sparingly.

This idea is purported in my mind from the International Mine Water Association (IMWA) annual conference held in Leipzig, Germany, in 2016 wherein special attention is given on the topic of "extracting value from mine water" and lectures were invited from world over. It is understood that it will be a rare situation in which extracted values can pay for all of the costs of water treatment; however, it would be good to extract value from the mine water and this can partially defray the costs of water treatment and both short-term and long-term gain can be made.

Nearly all mines whether surface or underground are situated in far-flung *plain areas* or *hilly areas* with single-layer and multiple-layer aquifers. These, surface mine pits, operational or mined-out pit, from where minerals are extracted, contain huge quantity of mine water (**Table 6** column 4). The extraction of value from the water of mine can be done by the following uses:

a. Coal quality improvement by coal washing at the pithead washery or in an installation closer to the mine

b. For ore cleaning and in metallurgical process

c. For construction-related civil works in mines/plants

d. For haul road wetting to suppress dust

e. Irrigation/agricultural uses of mine water after quality analysis

f. For miscellaneous uses by mine/plant colonies, gardening, etc.

The value addition from the mine water can be easily and effectively implemented into practice through corporate social responsibility (CSR) scheme of the mining company concerned. In the mines and mining industry, corporate social responsibility assumes a highlighted importance. Some noted examples of value addition are:

1. Putki-Ballihari colliery of Bharat Coking Coal Limited (BCCL) in India has 400 L of mine water treatment pilot plant for miscellaneous uses by villagers which help in extracting value from mine water and water conservation both.

2. In the course of coal mining, Western Coalfields Limited (WCL) mines tap a huge quantity of underground mine water (basically groundwater) out of which a small portion is utilized for its day-to-day functions, namely, coal washing, dust suppression in transport routes, domestic uses, etc. (**Table 6**). A major quantum of water went unutilized. Now these practices are being changed slowly, and water is being utilized for various developmental activities in nearby areas, for example, recharging canals, wells, and rivers, providing water for irrigation, and providing potable water to local people. WCL is using water from mines to help people combat water scarcity in and around mining areas. In many mining regions, thus converting mine water into potable water is one useful value addition by which number of mines can be benefited. Such initiative also helps mining company to develop strong societal bond between mine management and local population.

A packaged drinking water plant (RO plant of Coal Neer) at Patansaongi mine is yet another example of extracting values from the mine water (see facing figure) by the coal mining company WCL.

3. In one of the cement plant that uses coal for firing cement kiln, a thought to pre-wash coal using mine water was attempted to get the benefit of less coal consumption and low ash generation at source. The operational efficiency of kiln/plant can thus be enhanced considerably with less coal consumption and more energy use. Various captive mines of cement plant, owned by private companies, can get commercial benefits from this idea by making effective use of mine water within their industrial areas and for a selected specific purpose.

Some other ways to extract values from water available in mine are:

i. Both treated/non-treated mine water can be supplied/sold by company at a cost for the miscellaneous uses to local users. In an arid region, for example, Rajasthan, India, this may bring great value.

S. no	Name of coal mine	Purpose	Details	Remarks
(1)	(2)	(3)	(4)	(5)
1.	Saoner Underground Mine	Recharge of water channels/ rivulet and improvement in water level	Not available	For miscellaneous uses. Boregaon village is the beneficiary
2.	Bhatadi Mine	Recharging of water table	5.66 million LPD	Reuse of mine water in Chandrapur district
3.	Padmapur Mine	Recharging of water table and reuse of mine water	2.46 million LPD	Chandrapur district
4.	Kamptee O/C Mine	To provide potable water to Kanhan township from the mine water discharge. Also for irrigation uses	Discharge of mine water; 47 lakhs; 3.5 km pipeline from mine to filter plant	"K2K": Kamptee to Kanhan Project, Oct 2015 Water filtration plant in Kanhan Municipal area; 800 people benefitted
5.	AB Incline, Pipla and Silewara Mine	For irrigation uses by pipelines	Mine water discharge	Recharging of canal system for agriculture (2 km long); 350 farmer families benefitted
6.	Adasa Underground Mine	Mine water for irrigation	—	Angewada and patkakhedi villages; recharge of water table
7.	Bhanegaon Opencast Mine	For recharge of wells and river recharging	Uses in irrigation and agriculture for cultivation	Bina and Bhanegaon villages
8.	Murpar Underground Mine	Pumped out for supply to village pond	3500 GPM water; 50 ha land irrigated	Around 2000 population benefited
9.	Wani North	Recharge of water table through groundwater augmentation, that is, desilting and deepening of water bodies and villages ponds, etc.	48 lakhs; under CSR activities of company	Jalyukt Abhiyan of the Maharashtra Government for Nagpur and Chandrapur district in India

Source: WCL, Nagpur

Table 6.
Mine water uses at WCL coal mines: extracting value

ii. In captive mines of nearly all companies/cement plants, etc., "water charges" are paid to the government by the industrial organizations to make use of fresh groundwater for their industrial operations, for example, cooling, colony supply and other miscellaneous uses, etc. The fresh water use can be replaced with mine water, and a considerable value addition from mine water is possible.

iii. Immense value can be extracted, had underground mine water is made as potable. If treated using cost-effective techniques, conferring to the drinking water quality standards [38, 39], the potable mine water can also help in removing water crisis of the area too.

In brief, to extract the value from the mine water, a number of novel and innovative ideas are available, but practical methods as per local requirement are extremely useful. The abovementioned examples are simply a curtain raiser and not to be seen as the ultimate for value extraction from the mine water.

8. Groundwater and mine water management

Groundwater is a resource first. Its management at mine level and as mine water is a major challenge. To preserve and protect groundwater in mining areas from overexploitation and to manage it properly, the approach should be site-specific and engineering-oriented. Mining industry worldwide manages it through dewatering/pumping economically and properly. Both quality and quantity of groundwater and the intricate relationships between physical, chemical, and biological processes within mine deposits are a key to the development of effective strategies for water management. Application of "preventive approach" along with artificial recharge techniques and water conservation measures can remove the water management bottlenecks to a large extent. Hence, an approach to deal groundwater effectively works out better at different levels of management. The entire water management chain should be understood by all levels of management, that is, at corporate, at site, and at operations management level, and it has to be a "bottom-up approach rather than a top-down approach."

Looking at operational stage of surface mines (pit mine) which are working below the water table, the puncturing of water table results into the accumulation of water on dip side of the open mines. Due to heavy precipitation in limited period (downpour), such water accumulation problems lead to the hampering of normal mine production. Similarly watery underground mines have multistage pumping needs. Sustainable water management at mine sites has close linkages with production; hence to improve water management in the mining environment, the following areas need attention:

1. Corporate policy with respect to water management

2. Planning and machinery used, that is, pumps and pumping

3. Mine drainage

4. Water quantity

5. Water quality and pollution

6. Ore processing, tailings, and waste disposal

Clear and transparent policy and sound water management give a direction to the implementation plan and programs economically and as per the desire of the organization. Improvement in water management practices periodically and practically is imperative for GW management.

Planning and machinery used involve the site conditions and stage of operations, in the chain of water management. It requires innovative thinking so that planning is practical in implementation and percentage utilization of machinery is maximum. Routine condition monitoring (RCM) for routine maintenance of equipment and machinery should be cost-effective for proper water management in general.

Mine drainage can pose a serious threat to water quality and mine productivity. The importance of this issue becomes more critical as demand for resources grow. When complex metallic ore deposits are mined, the geochemical evaluation of mine drainage water becomes important in pollution evaluation as well as deciding prevention strategies. Their economical remediation is possible to an extent through proper mine drainage system.

Undoubtedly, in ore processing, tailings, and waste disposal, methods and procedures are key areas of focus in the pollution abatement strategies. Therefore, to deal with it, attention toward the "best practices in water management" is needed. Practically, for improving water management in the mining environments, approaches should be:

i. Long-term and short-term costs of water treatment: Mine water management involves its treatment too. The mine water and its treatment involve a sizable long-term and short-term costs of water treatment. If the cost economics are understood correctly, it can be applied in curbing the overall cost of mineral production. Mine water is valuable in terms of its quantity and ease in extracting it for bulk reuse, if available in open-pit mine. The "pit water" may or may not confer to the prescribed quality standards, either for irrigational or miscellaneous uses, but certainly able to meet out the water scarcity in an area. Therefore, its treatment is sought after, though considered as not economical. In the case of underground mines, how the mine water increases overall cost of mineral production (**Box 2**) needs to be understood?

ii. Local solutions are always cost-effective in mining activity because most of the time it requires crude solutions (and not very precise) on site-specific basis.

Box 2. How mine water increases cost of production?

Mine water presence in excess requires pre-draining that adds to the cost of pumping; more expensive construction prevents the use of preferred methods and equipments. Overall it puts additional burden on cost of mineral production. If underground mine is watery, it requires use of more expensive explosive. "Timber support," if used in underground mines, are not good if mine has wet conditions. Alternate wetting and drying of timber cause timber decay and endanger mine safety. The mine water washes weak ground from underground openings, for example, sand, silt, gravel, clay, etc. are washed easily causing reduced safety for wall, roof, etc.

In the case of underground mine water, if one knows the effect of water on surrounding underground environment, its value can be assessed both directly and indirectly, for example, water is hazardous in mine shaft because wetness corrodes hoist ropes, steel girders, ladders, planks, shaft timber, etc.; mine water and wetness add to the maintenance of underground equipments, reduce effectiveness of lubricant, increase corrosion, cause scaling in pipes, lead to rusting in wet exposed metallic surfaces, etc.; mine water may add to miner's discomfort due to continuous wetness of protective clothes and bring illness (a form of indirect cost). Increased electrical hazards are the anticipated effect of mine water on mine safety underground.

iii. Preventive water management (PWM): impact of mining on groundwater and its imprints should be kept controlled. Through this approach and in the mine catchment area, this can be done by preventive water management. Plans for PWM should be such so that attention is paid to both quality and quantity aspects as they are to be managed with the ultimate goal of achieving an ecological balance. Some key points as regards with this approach are:

- Preventive approach has the ability to remove or add the nutrients from soil/land (through surface water) because land, soil, and water are an integrated part of natural water system. By this approach soil/land quality can be made sustainable, for example, (a) nitrogen compounds are broken down and phosphate is fixed for agricultural land use in plain profile land areas with adequate water. (b) By allocating proper land use profiles with the land use activities for each catchment area, an improvement in overall land quality is achievable. (c) Designing of suitable land use pattern within the mine lease area or catchment areas is a step forward toward mitigation and preventive care.

- Adequate drainage pattern of mine/project area commensurate with the natural drainage.

- Effective local runoff arrangement of rainwater for GW infiltration into the soil and in hard-rock areas.

- Widening of watercourses, cleaning of silt from pond/tank beds or open ditches and raising the drainage level of water channels thereby increasing groundwater recharge and the water storage capacity.

The description given above explains that how management of water can be done economically and effectively with practically implementable water management practices. Since underground mine water management is sharply different from open-pit water management, technical knowledge of mining engineering can be an added advantage. However, it would be good to extract value from the mine water, as doing so can partially recover the water treatment costs and both short-term and long-term gain can be made. It is also understood that proper water treatment and management with respect to mines can bring a stage/situation in which groundwater will turn into a useful commodity for that particular mine which is scientifically managed and evaluated.

8.1 Solutions for water problem in mines

In mining sector, which is prospective, very large, and capital intensive too, scientific approach toward groundwater management should be applied to curb and restrict groundwater overexploitation and maintain basic groundwater equation, that is, more recharge, less draft. To tackle water problem in both mine types, the following solutions are noteworthy:

- Best management practice (BMP): Number of solutions for open-pit as well as underground mines can be solved through case study experiences available internationally. Some *best management practice* (BMP) for water in mining and mine environment area are important from this view point [40, 41]. Guidance manual and case study experiences in various parts of the industry worldwide, provide solutions together with lessons (to be learnt) for better understanding.

Sometimes BMP is also referred as "best practice mining" or "best mining practice." In brief, BMP does not refer to any designed/formulated method but implies to "the continuous improvement of mining and management practices to maintain maximum performance for achieving an acceptable level of environmental protection." In doing so, it is necessary to incorporate and integrate economic, environmental, and social considerations into the mining operations in a practical way.

Mining involves mostly excavation, loading, and transportation operations. The most environment-unfriendly among these is the "transportation" and "dust generation by transportation." By adopting BMP the stress on environment is reduced because BMP emphasizes curtailing unscientific practices and avoiding shortcuts. Effective surface water utilization is the best management practice (BMP) for optimum use of rain-fed water resources. Similarly, pollution control measure as applicable to large-sized public sector mines, that is, preventive approach, control at source, and zero liquid discharge (ZLD), is a solution through BMP.

- Integrated water resource management (IWRM): This becomes relevant when addressing water availability, water security, and water access for all users. IWRM involves the coordination of stakeholders in the water use of a site, an area or region to ensure economic and social development together with maintaining the ecosystem balance. Based on IWRM and stakeholders' experiences, water policy can be made sound, and balanced decisions in response to specific water challenges, being faced by the industrial company, can be taken. It is always desirable that cooperation between community, authorities, and organizations be maintained and public participation in water management be encouraged.

  Thus, IWRM is an interdisciplinary approach to devise and implement efficient, equitable, and sustainable solutions to water and development problems. This approach is open and flexible and brings together decision-makers across the various sectors that impact water resources. IWRM principle ensures that water is sufficient for industrial operation and all users too. These days companies are concerned about continued water access in light of increasing scarcity. Their response is to maximize their efficiencies and limit their inputs. IWRM also involves "standardized water reporting," which is a low priority issue for the operating mines or industry. The issue with water reporting is that of hiding impacts of mine water-related issues with communities and regulatory authorities. Money/financial obligations are the principle cause for this hiding. Beyond this there exists a need to comply national environmental laws/regulations, which should be complied and put into practice. Some of the barriers to IWRM in the past were the lack of hydrological data and models which have been overcome these days by the scientific studies. IWRM together with BMP (best mining and management practices) is capable to yield desired water resource management results as expected.

- Sustainable groundwater exploitation policy: Area/states which are mineral-rich and having number of operative mines should formulate and frame a sustainable groundwater exploitation policy for mines separately in line with the "National Water Policy." For groundwater protection, practically applicable regulatory framework [42] should be in place and enforced strictly for solving water problems.

- Reliable mine water technology: Open-pit mine water management contains number of lacunas, and these can be reduced by bridging the gaps in water use and reuse, for example, surface mining operations in water-stressed and water-critical areas. Since maximum mineral production is achieved from surface mines, industrial attempt should be such so that reduced water consumption philosophy be adopted for excavation and ancillary industrial operations. This also makes mine water sufficient. Reliable mine water technology [43] is yet another option for tacking mine water problem.

- Recycling, conservation, and recharge: Promotion and encouragement for mine water recycling/reuse, water conservation, and groundwater recharge can remove water crisis in and around the mining site. In this regard, sub-categorization of water as "surface water" and "groundwater" will provide better solution. By addressing the impressive technical solutions related to water pollution, positive results can be achieved.

To curb the overexploitation (excessive withdrawal of groundwater from aquifer) for industrial purpose, imposition of tax or *cess* and pricing of the groundwater use is a way out. To conserve groundwater and rationalize the groundwater use by the mines, limited withdrawal permission is helpful in the excessive groundwater exploitation.

Rainwater harvesting (RWH), the most popular method of groundwater recharge, is the best solution to reduce dependence on groundwater. Implementation of these techniques and optimization of innovative alternates of RWH need to be encouraged according to the mine needs and requirements to provide solution locally.

- U/G versus surface mine: Underground (U/G) mines and surface mine's water-related problems are different. Therefore, solution to tackle water problem in underground mines are also typically different. Some encountered conditions of underground mines are:

  Condition (i): sudden inrush of water or heavy water seepage from surface water body to underground mine workings in proximity, leading to inundation
  Condition (ii): underground mining near "perched water table" (an accumulated/stored water underground)
  Condition (iii): unprecedented or accidental connection of underground mine workings with aquifer containing infinite amount of water or water under pressurized conditions

Underground mines either operative or abandoned when filled with water pose a problem of "mine inundation." Many times such inundated waterlogged areas lead to mine disasters and also hamper normal mineral production in underground mines. The worst ever disaster caused by mine inundation in India was at "Chasnala Colliery" in the state of Bihar, India, in the year 1975 wherein 372 persons were drowned underground. Underground galleries approached the waterlogged old workings of an abandoned mine and faulty prediction of mine development had caused this accident to happen. The safest procedure to deal with inundation in mines is never to take the position of old working "for granted" until they have actually been proved by proper survey. No mine working which has approached within a distance of 60 m of any disused or abandoned working, whether in the same mine or in an adjoining mine, shall be extended further to endanger safety.

In underground mining, the mining operation near water bodies [44] assumes significant importance from research point of view. This is principally due to the uncertainty involved. Behavior of the surface water bodies (water head), intervening strata over the mine workings, its location (in the buffer zone/core zone of the mine lease area), and in between distances, plays considerable role and hence assumes significant importance. Therefore, geological, mining, and hydrological parameters must be looked while evaluating the real field situation, for example, topographical features such as hills/valley(ies) or ravines land, etc. should be considered. For solutions one must observe, examine, and check the proximity of old underground mine workings, whether the area is dry/damp or seeping in (heavy/low). It is possible that the workings of adjacent mine may not be filled with water but the barrier pillar and its thickness are important and must be maintained as per the statutory requirement or the existing guidelines framed for the purpose.

To search solutions for water problem of underground mines, due consideration should be given for water impoundments (stagnant water bodies) on surface as well. Seasonal or perennial streams, standing water bodies, and sea vicinity to the mine are important for pressure head created by the surface water or impoundments. Similarly, underground mining should not come closer to active oil and gas well (150 ft. minimum).

9. Conclusions

Groundwater during mining of minerals causes problems related to environmental impact, most commonly *depletion of water tables* in and around mining areas, which in turn leads to social/industrial unrest. Hence, ground water quantity pumped out of the mine should not be more and known as well, for proper use. Every necessary effort shall be made to restore original level of water table below ground, whether it is open-pit mining operation or underground mining or any other similar excavation. In brief, and summarily, it is inferred that mining industry should guide their efforts toward proper use of groundwater encountered in mines and avoid wastages because water quantity handled in mines is very large.

Mine, being an important mineral production enterprise and groundwater as a valuable resource being continually under stress, has to be assessed scientifically from industrial perspective. The water management measures shall be identified beforehand and remedial measures be kept in place. To augment water level, artificial recharge of groundwater by rainwater harvesting, creation of pit lakes/water lagoons, and recharge through abandoned tube wells are some easy and economical measures. Needless to say the basic principle of sustainability, that is, conserving for future generation, must be adhered.

Author details

Abhay Kumar Soni
CSIR—Central Institute of Mining and Fuel Research (CSIR—CIMFR), Nagpur, Maharashtra, India

*Address all correspondence to: abhayksoni@gmail.com

References

[1] Gleeson T, Befus KM, Jasechko S, Luijendijk E, Cardenas MB. The global volume and distribution of modern groundwater. Nature Geo-science. 2016; **9**:161-170. DOI: 10.1038/NGEO2590. Available from: www.nature.com/naturegeoscience [published online on 16 Nov. 2015]

[2] UNICEF [Internet]. Water in India—Situation and, United Nation's India Children Education Fund (UNICEF). New Delhi; 2013. p. 93. Available from: http://www.indiaenvironmentportal.org.in/files/file/water%20in%20india.pdf [Accessed: 05-2014, 2017 & 2018]

[3] Ayoob S, Gupta AK. Fluoride in drinking water: A review on the status and stress effects. Critical Reviews in Environmental Science and Technology. 2006;**36**(6):433-487

[4] Jacks G, Bhattacharya P, Chuodhary V, Singh KP. Control on genesis on some high-fluoride groundwater in India. Applied Geochemistry. 2005;**20**:221-228

[5] Dharashivkar AP, Khond MV, Murkute YA. Preliminary report of fluoride contamination in groundwater from PG-2 watershed of Penganga River Basin, Korpana Taluka, Chandrapur District, Maharashtra. Gondwana Geological Magazine. 2014;**14**:161-166

[6] Waghmare SS, Tanvir A. Fluoride removal from water by various techniques: A review. International Journal of Innovative Science Engineering and Technology (IJISET). 2015;**2**:9. ISSN: s2348-7968. Available from: www.ijiset.com

[7] Available from CPCB web page: http://cpcb.nic.in/industry-effluent-standards/

[8] Wood Peter A. Hydrological problems of surface mining, Report No. ICTIS/TR17. London: IEA Coal Research; 1981. p. 102

[9] Maitra MK. Watershed Management—Project Planning, Development and Implementation. New Delhi: Omega Scientific Publisher; 2001. p. 216. ISBN: 8185399549

[10] Lamare RE, Singh OP. Limestone mining and its environmental implications in Meghalaya, India. ENVIS Bulletin Himalayan Ecology. 2016;**24**:87-100

[11] Lamare RE, Singh OP. Degradation in water quality due to limestone mining in East Jaintia Hills, Meghalaya, India. International Research Journal of Environmental Science. 2014;**3**(05): 13-20

[12] Lamare RE, Singh OP. Localized effect of artisanal and small scale mining of limestone mining on water quality in Meghalaya, India. Pollution Research. 2015;**34**(02):321-329

[13] Canter LW. Environmental Impact Assessment. Vol. 660. Singapore: McGraw Hill International Editions; 1996

[14] USDOI: Groundwater Manual—A Water Resources Technical Publication, US Department of the Interior (USDOI), Indian Reprint. Jodhpur: Scientific Publisher; 1985. 480 p

[15] Tarnekar PS. Study of environmental implications of water quality with special reference to geology and mining activities in parts of Nagpur and Bhandara districts of Maharashtra [Ph.D. thesis, unpublished]. Nagpur: Rashtrasant Tukadoji Maharaj Nagpur University (RTMNU); 1993

[16] Soni AK. Groundwater abstraction in mining for industrial applications:

Obtaining permission for statutory compliance. Mining Engineering Journal. 2018;**12**:22-26

[17] Chunhu Z, Jin D, Geng J. Numerical simulation of the groundwater system for mining shallow buried coal seams in the ecologically fragile areas of Western China. Mine Water Environment. 2019; **38**(1):158-165. DOI: 10.1007/s10230-018-0551-z. (Now published ; First online in 2018)

[18] Abbi YP, Jain S. Handbook on Engineering Audit and Environmental Management. New Delhi: The Energy Research Institute (TERI) Press; 2006. pp. 148-164

[19] Choubey VD. Hydrological and environmental impact of coal mining, Jharia Coalfield, India. Environmental Geology. 1991;**17**:185-194

[20] Allen SK, Allen JM, Lucas S. Concentration of containments in surface water samples collected in west-central impacted by acid mine drainage. Environmental Geology. 1996;**27**:34-37

[21] Tiwary RK. Environmental impact of coal mining on water regime and its management. Water, Air and Soil Pollution. 2001;**132**(1–2):185-199

[22] Khan R, Israili SH, Ahmad H, Mohan A. Heavy metal pollution assessment in surface water bodies and its suitability for irrigation around the Neyveli lignite mines and associated industrial complex, Tamil Nadu, India. Mine Water and the Environment. 2005;**24**:155-161

[23] Soni AK. Evaluation of hydro-geological parameters associated with limestone mining: A case study from Chandrapur India. Mine Water Environment. 2007;**26**:110-118

[24] CGWB. Groundwater Estimation Methodology (GEC). New Delhi: Central Groundwater Board (CGWB), Government of India; 1997. Available from: http://cgwb.gov.in/

[25] NABARD. Review of Methodologies for Estimation of Groundwater Resources in India. New Delhi: National Bank for Agriculture and Rural Development (NABARD); 2006. 179 p

[26] CSIR-CIMFR. Geo-hydrological Study for Mining at Deeper Levels in Limestone Mine of M/s Trinetra Cement, District— Banswara, Rajasthan, India (No.: CNP/N/3259/2012-13). Dhanbad: Central Institute of Mining and Fuel Research (CIMFR); 2013. 73 p

[27] CSIR-CMRI. Comprehensive Geo-hydrological Studies for Rajhara Mechanized Mine of Bhilai Steel Plant (BSP), Dalli Rajhara, District—Durg (Chhattisgarh), No. : GC/MT/N/22/2005-2006. Dhanbad: Central Mining Research Institute (CMRI); 2007. 168 p

[28] CSIR-CMRI. Hydro-geological Study of Malanjkhand Copper Project, Hindustan Copper Limited, District Balaghat (M.P.), No. SSP/N/235/2017-18. Dhanbad: Central Mining Research Institute (CMRI); 2018. 114 p

[29] CSIR-CIMFR. Geo-hydrological Studies for Quantitative Estimation of Water and its Management at the Lanjiberna Limestone and Dolomite Mine of M/S OCL Limited, District-Sundergarh, Odisha (No.: CNP/N/2344/2009-2010). Dhanbad: Central Institute of Mining and Fuel Research (CIMFR); 2011. 98 p

[30] Soni AK, Sahoo LK, Ghosh UK. Quantitative estimation of water in an opencast limestone mine of India. The Indian Mineral Industry Journals. 2016; 21-29(11) (also published in Geominetech Symposium on New Equipment, New Technology, Management and Safety in Mines and

Mineral Based Industries (ENTMS); 11-12 May; Bhubaneshwar)

[31] Rawat NS, Viswanathan S. Assessment of water quality deterioration in some coal mines of Northeastern Coalfields and Jharia Coalfields. Journal of Mines, Metals & Fuel. 1990;**01**:15-21

[32] Mohabey NK, Taranekar PS, Thergaonkar VP. Impact of mining activity on water quality of Umrer Coalfield. Journal of Gondwana Geological Society. Golden Jubilee Special Volume. 1996:303-315

[33] Singh AK, Mahato MK, Neogi B, Singh KK. Quality assessment of mine water in the Raniganj Coalfield Area, India. Mine Water Environment. 2010; **29**:248-262

[34] Handa BK. Water Sampling and Analysis, Chapter 4, New Delhi: Govt. of India; 1974. pp. 51-72. Available from: http://www.who.int/water-sanitaion-health/dwg/2ed vol3d.pdf

[35] APHA. Standard Methods for the Examination of Water and Waste Water. 14th ed. Washington, DC: American Public Health Association (APHA); 1998. pp. 6-187

[36] Khond MV. Hydro-geochemical characterization of groundwater around Naokari and Manikgarh Limestone mines, Chandrapur District, Maharashtra [Ph.D. thesis unpublished]. Nagpur: Rashtrasant Tukadoji Maharaj Nagpur University (RTMNU); 2014

[37] Soni AK, Wolkersdorfer C. Mine water: Policy perspective for improving water management in the mining environment with respect to developing economies. International Journal of Mining, Reclamation and Environment. 2015;**30**(2):115-127. DOI: 10.1080/17480930.2015.1011372

[38] BIS, Indian Standard. Specification for Drinking Water (IS:10500). New Delhi: Bureau of Indian Standard (BIS); 2003. pp. 1-5

[39] WHO. Guideline for Drinking Water Quality Recommendation. World Health Organization (WHO). 2004;**1**:1-4

[40] EPA. Overview of Best Practice Environmental Management in Mining, Australia. Environmental Protection Agency (EPA); 2002

[41] USEPA. Coal Re-mining Best Management Practices Guidance Manual. Washington, DC, USA: U.S. Environmental Protection Agency; 2001

[42] CGWA. Draft Guidelines for Issuance of No Objection Certificate (NOC) for Groundwater Withdrawal. Central Groundwater Authority (CGWA), Government of India; 2017. Available from: http://www.cgwa-noc.gov.in/Sub/CheckEligibility/Check Eligibility.aspx

[43] Brown A. Reliable mine water technology. Journal of Mine Water and the Environment. 2010;**29**(2):85-91

[44] Lele MG. Water Problems in Mines. 1st ed. Institute of Technology—BHU Varanasi. India: Nishkam Press; 1980. 399 p

Chapter 6

Groundwater Protection Legislation in Slovenia: Theory and Practice

Miha Curk and Matjaž Glavan

Abstract

As an EU member, Slovenia implemented the Water Framework Directive (WFD) guidelines into national legislation quite successfully, but in many parts of the country, groundwater is still under threat, mainly from nitrate. The problematic areas, as is the case in many other EU countries, are mostly shallow-soiled alluvial plains. Their groundwater is the country's biggest source of drinking water, but at the same time, the fertile soil on their flat surface is considered to be the most suitable for agricultural activities. We are aiming to provide an overview of groundwater protection practices in Slovenia. To evaluate the "theory," we will take a close look at the national legislation concerning the subject. From the "practical" perspective, we will research what guidelines and solutions were drawn from legislation to comply with WFD objectives. Furthermore, we will also discuss the current activities aimed at improving Slovenia's groundwater status.

Keywords: groundwater protection, legislation, Water Framework Directive, nitrate, agriculture, best management practices, shallow soil, alluvial plain, drinking water

1. Introduction

Groundwater in Slovenia, as well as in many places across the globe, is the main source of drinking water. Apart from its use for drinking, it is often used for irrigation and other purposes. This makes groundwater valuable, and since it generally does not renew very quickly, it is often—as is the case in Slovenia—protected by national legislation. Quantity, however, is not the only issue we are facing in regards to groundwater. With the rapid increase in use of mineral nitrogen fertilizers and different agricultural as well as industrial chemicals in the past century, many of these substances found their way into groundwater. In last few decades since the pollution was becoming more and more severe and even health threatening—a lot was done in terms of improving groundwater protection. Slovenia as an European Union member formed its legislation based on Water Framework Directive (WFD) and its recommendations. But despite that, groundwater status in some parts of the country is still not adequate, mostly because of the presence of nitrate. Main vulnerable areas are large alluvial plains—their flatness makes them very suitable for agricultural production, but their shallow soils make almost every activity on top of them affect the groundwater quality. The struggle is real, and on the following pages, we are about to uncover the theory and practice behind the groundwater protection in Slovenia.

2. Groundwater pollution in Slovenia—the background

Slovenia is not a big country at just over 20,000 km^2. It is, however, very diverse in terrain and climate. In north-west, it touches the Alps, in south-west, the Adriatic Sea, and in north-east, the Pannonian Basin. Except for the latter part, most of its relief consists of hills and mountains, and the only larger flat areas are alluvial plains, whose aquifers serve as storage for the majority of country's groundwater. Climate follows the characteristics of terrain, which means that the western parts of the country get 2,000 and up to 3,000 mm, while the north-east can get as little as 800 mm of rain in a year. The abundance of rain is the main reason for Slovenia's groundwater capacities, but the seasonal variability in recent years is not best suited for agriculture, as peaks are moving more and more into autumn, while in other seasons, rainfall is diminishing [1].

As mentioned above, main reservoirs of groundwater are located under alluvial plains. These are also centers of agricultural production, with intensive arable rotations and vegetable fields, but in some areas also orchards. According to National Environmental Agency (ARSO), the main problematic areas are Murska kotlina, Dravska kotlina, Savinjska kotlina, and Krška kotlina, labeled in top to bottom order in **Figure 1** with an exclamation mark sign. It is quite easy to notice that they share a similar type of an aquifer, and the depth from ground surface to it is often times as shallow as 1–2 m. Views from Murska kotlina and Krško polje can be seen in (**Figures 2** and **3**).

Speaking of pollution, there are various different substances causing it. As mentioned before, nitrate is one, but certainly not the only troublemaker. Pesticides can also pose significant threat, and in Slovenia, atrazine is the most common of them. In its reports, ARSO also mentions other pesticides, like bentazone, terbutilazine, isoproturon, chloridazon, etc., but these are mostly just detected, and do not pose

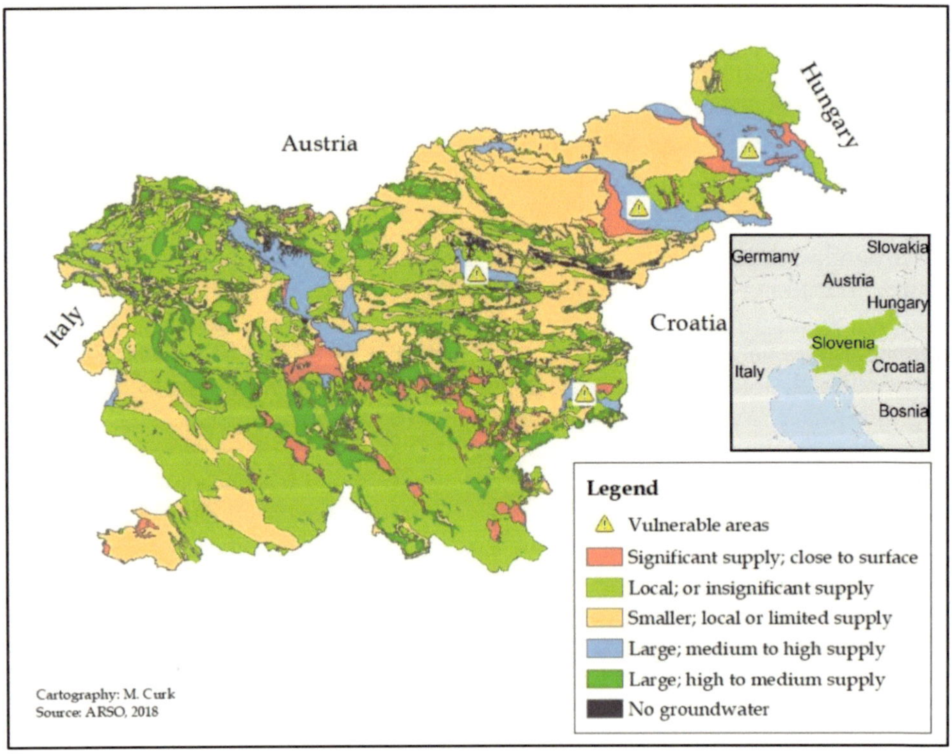

Figure 1.
Map of Slovenia with different aquifer types; blue color mostly corresponds to the main alluvial plains.

Figure 2.
Corn harvest on Murska kotlina plain (photo: Curk, 2017).

any threat. Atrazine and its metabolite desetil-atrazine are still found in quite high concentrations though, in places well above the EU threshold of 0.1 μg/L. What is interesting is that, despite atrazine being banned in EU (or rather its registration was not renewed) more than 15 years ago, it is still present in our groundwater (**Figure 2**).

3. Processes behind the pollution—a word on leaching

When speaking of groundwater pollution, leaching is generally the cause of it. Molecules that do not have the positive charge and therefore do not bind well to soil particles get washed through the soil profile as water passes through. Nitrate and many other substances have this exact property which makes them very easy to leach into groundwater. We will only talk about nitrate in the following lines, as it represents the process very well, and is common across the globe, unlike specific pesticides.

Nitrogen as an element is an essential component of plants, so in order for them to grow, we need to supply it to them. For centuries, this was done only with manures, but after the green revolution in the last century, a new source of nitrogen was introduced in form of mineral fertilizers. These allowed for a great increase in agricultural yields (up to 40%), but all the nitrogen from fertilizing did not make it into crops. Because of aforementioned leaching, much of it was lost into ground (or surface) water in the form of nitrate (NO_3^-). Fertilizers with other forms of nitrogen, like ammonia (NH_4^+) or nitrite (NO_2^-), are not much "safer" in this regard, as nitrogen forms are prone to quickly transform in nitrification and denitrification reactions. There was a debate in recent years about nitrification inhibitors, compounds used for decreasing of nitrate leaching by inhibiting conversion of ammonium into nitrate. Ammonium, having a positive charge, bounds to soil particles and is not as prone to leaching. But studies have shown that decreasing of nitrification results in ammonia volatilization increase, which, environmentally, is not any better [2, 3].

Some areas, however, have less trouble with leaching than others. As mentioned, leaching is strongly dependent on soil type, climate, and agricultural practice. Here are some examples: sandy soils that are not as good in water retention are notorious for leaching problems, while clayey soils are better in this regard; water and nitrate are held in root zone for longer, so plants can use more of them. Climate wise, the more problematic are areas with stormy weather and lots of rain during the spring to early summer period, when most of the fertilizer is applied; and thirdly, agrotechnical

Figure 3.
A view of agricultural land on Krško polje plain (photo: Curk, 2018).

practices, which are the only piece of the puzzle we can actually control, also influence leaching significantly. Fertilizing only once per year is way worse than in several small rations, where plants get as much as they need when they need it. Plowing fields in autumn is worse than in spring because winter rains do not leech the nutrients from decaying organic matter we buried in the soil. It is good to use cover crops when soil would otherwise be left bare and use crops with deeper root systems that catch nitrate in deeper layers as well [4–6]. Having all the best factors come together in one area is not realistic though. It is always necessary to compromise—if we have good soil and climate conditions, we could be a bit more "sloppy" with our agrotechnics and not get alarming monitoring results, but if soil and climate are not the most suitable, even the best agrotechnics might result in noticeable leaching. What is important in such cases is dialog between farming and water protection sectors to solve the dilemma of enough food versus clean environment. Most of the nitrate and pesticide problems could probably be alleviated by banning agriculture from vulnerable areas, like sandy shallow-soiled plains above drinking water reservoirs, and environmental side often tries to suggest this option. But in reality, such areas are many times the only areas suitable for intensive agriculture. This is why we adopted threshold values, which are a consensus between health and other benefits. European Union Water Framework Directive therefore recommends a nitrate threshold of 50 mg/L for water used for human consumption, because this concentration is low enough to ensure safe consumption, but high enough that it still allows for use of agricultural land (**Figure 3**).

4. Slovenian water protection legislation—the theory

Slovenia is not a very old state. For centuries, it was a part of different Germanic kingdoms like Austro-Hungarian Empire, but in twentieth century, it mostly existed as a state in different connections of South Slavic federations (most recently in the socialistic Yugoslavia). In 1991, Slovenians finally managed to form our own state, and in 2004, we decided to join the European Union. Since the secession from Yugoslavia, a lot of legislation was taken over from Yugoslavian, but in terms of water protection, major improvements were done since the accession negotiations with European Union (**Figure 4**).

The main piece of legislation concerning the subject is the Water Framework Directive (Directive 2000/60/EC of the European Parliament and of the Council)

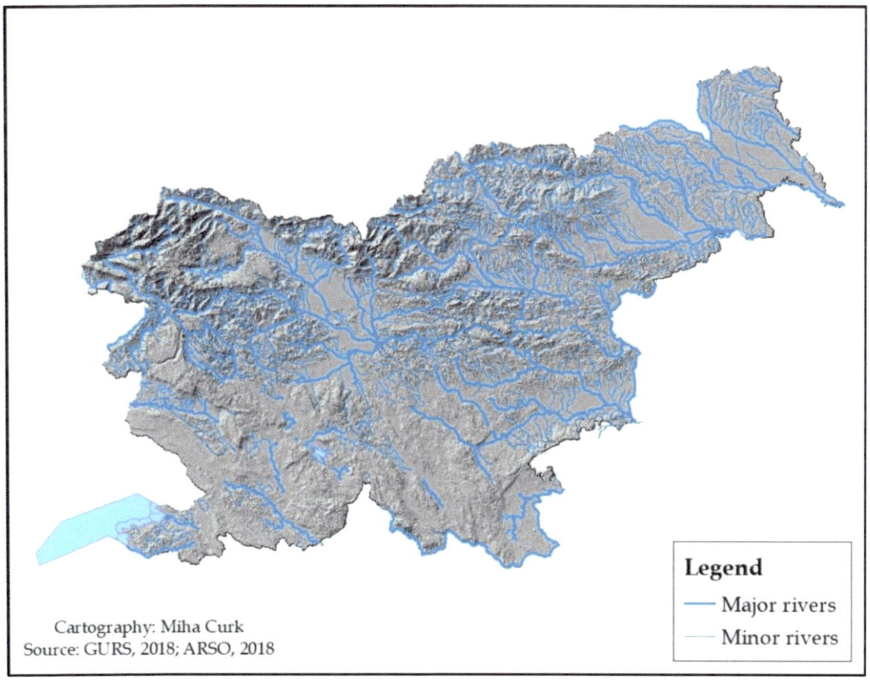

Figure 4.
Surface water and topography in Slovenia.

in force since October 23, 2000. Its main objective was establishing a framework for Community action in the field of water policy. Member states had to implement the necessary measures to prevent deterioration of the status of all bodies of surface water (art. 4.1(a)(i)); and protect, enhance, and restore all bodies of surface water to achieve good water status (art. 4.1(a)(ii)).

In Slovenia, Water Framework Directive resonated in acceptance of Waters Act [7]. Waters Act defines rules for management of waters (both surface and subsurface) in Slovenia, specifically for protection of waters, sustainable use of waters, good management of water use facilities, and other water-related questions. Waters Act imposes the formulation of a Program of Measures which is used to define measures needed to meet the objectives set for protection, development, and use of waters. A state-wide Program of Measures needs to be formulated every 6 years by the government with the help of interested public. More locally specific programs should be formulated for river-basins or other vulnerable areas when necessary. Different daughter rules were also brought to act, for example:

- Rules on determining water bodies of groundwater [8] (define water bodies of groundwater based on aquifer location, groundwater movement, groundwater quality, and human activity above the water body).

- Rules on criteria for marking a water protection zone and a bathing water zone [9] (define means of designation for water protection zones—different warning signs, fencing, signs with instructions for acting in case of emergency).

- Rules on determining and classification for water bodies on surface water [10] (define surface water bodies depending on types of water body, hydromorphological changes, anthropogenic changes, and state of water).

They offer more specific limitations concerning the subjects. Specific water protection zones are defined by decrees that define locally specific areas around drinking water reservoirs, which are shown in **Figure 5**. Red areas mark the inner-most part of the zone directly around the pumping station. Strictest measures apply here. Transitioning from red across orange, and yellow to green, less and less strict measures apply. The size of each zone was determined by different factors; for example, by aquifer type, speed of groundwater recharge, response time, pollutant retention time, dilutive capacities of groundwater body, etc. Different modeling approaches were also used in the process to help determine groundwater flow, dilutive properties etc. A common criterion for definition is the time it takes for pollutants to reach the well. Time for the first zone is usually 50 days and second zone is 400 days, while the third zone encircles the whole connected aquifer.

Another important piece of legislation is the decree on groundwater status [11]. It was transposed from European Union Ground Water Directive with the aim to define procedures for determining of groundwater quality, standards of quality, and corresponding threshold values, but also other demands concerning preparation of groundwater protection action plans, etc. For any given groundwater body, investigation of pollution was executed, and bodies with good chemical status were determined. Bodies that were not recognized as such are considered vulnerable and a program of measures has to be formed with an aim to prevent further pollution and restore the body to good chemical state.

A groundwater body is considered to be of good chemical status when:

- chemical composition is such that yearly average of sampling locations results does not exceed threshold values.

- pollutant concentrations do not indicate intrusion of salt water or other intrusions into water body, do not indicate deterioration of connected surface water status, do not cause damage to aquatic or land ecosystems.

- electrical conductivity changes show that there are no intrusions of salt water or other substances into water body.

Threshold values for pollutants in groundwater are presented in **Table 1**. Note that values are similar to European Union Drinking water directive (Council Directive 98/83/EC of 3 November 1998 on the quality of water intended for human consumption). In Slovenia, as mentioned before, groundwater is used extensively for drinking (98% of drinking water comes from ground water), so EU drinking water values apply to all groundwater.

Apart from legislation that is more directly bound to groundwater issues, other legislation pieces are also important for our topic. Agriculture and industry influence groundwater quality significantly, so some relevant regulations also concern these sectors. Let us only concentrate on the agricultural part. For the most part, the framework for these regulations is the following European Union directives:

- Nitrates Directive (Council Directive 91/676/EEC of December 12, 1991 concerning the protection of waters against pollution caused by nitrates from agricultural sources).

- Pesticide Directive (Directive 2009/128/EC of the European Parliament and of the Council of October 21, 2009, establishing a framework for community action to achieve the sustainable use of pesticides).

- Sewage Sludge Directive (Council Directive 86/278/EEC of June 12, 1986 on the protection of the environment, and in particular of the soil, when sewage sludge is used in agriculture).

In Slovenia, Environmental Protection Act [12] is the framework act concerning protection of the environment. Its aim is to set up a framework of principles, measures, monitoring, etc., to ensure sustainable environment protection. It refers to other, more specific regulations, though, and these are listed below:

- Decree on the protection of waters against pollution caused by nitrates from agricultural sources [13] (below: nitrates decree)
- Plant protection products act [14] (below: PPP act)
- Decree on the management of sewage sludge from the urban waste water treatment plants [15] (below: sewage sludge decree)

Main emphases are collected in **Table 2**.

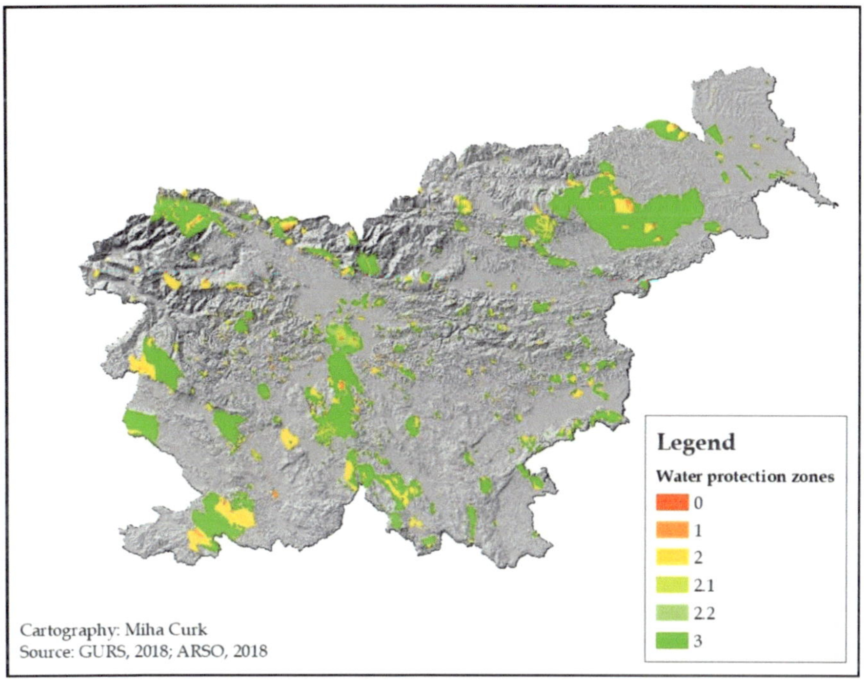

Figure 5.
Water protection zones in Slovenia—colors go from strictest (red) to least strict (dark green).

Pollutant	Threshold value	Unit
Nitrate	50	mg NO_3/L
Aldrin, dieldrin, heptachlor, heptachlor epoxide	0.03	μg/L
Other individual pesticides and their metabolites	0.12	μg/L
Sum of all detected pesticides and their metabolites	0.5	μg/L

Table 1.
Threshold values for pollutants from the decree on groundwater status [11].

Nitrates decree	Defines threshold values for agricultural nitrogen input into soil and water and measures for preventing water pollution. Yearly input of nitrogen should be less than 170 kg N/ha. Application of liquid manure is prohibited from November 15 till March 1 with some exceptions. Application of N containing mineral fertilizers is prohibited from October 15 till March 1. Any fertilization is prohibited on flooded, frozen, or snow covered ground regardless of these dates. Manure storage needs to be done in designated areas where leaching is limited. Crop rotation schedule needs to be prepared as a basis for well-balanced fertilizing.
PPP act	Defines means of proper use of plant protection products (PPP) in a way to allow for sustainable agriculture and safe environment for humans and animals. Also defines trading activities and their limitations. Only state-recognized PPP that are registered and published by responsible government department can be sold and used. PPP users need to undertake and pass a state-valid training program, special areas need to be designated for storage and mixing of PPP. Application equipment needs to be checked by responsible government department every 3 years. Treatment from air (i.e., with planes) is not allowed. Treatment close to streams or lakes is prohibited (15 m from the river bank).
Sewage sludge decree	Defines conditions and limitations of sewage sludge use in agriculture. Prescribes measures for processing of sludge before use, standards for processing, etc. Sewage sludge is considered problematic because of heavy metals. If a farm is involved in CAP payments, use of sewage sludge is prohibited.

Table 2.
Main emphases from nitrates decree, PPP act, and sewage sludge decree [13–15].

Apart from these state-wide valid limitations, special rules also apply to water protection zones. Three main zones are used (see **Figure 5** for reference) to divide land around groundwater reservoir depending on risk of harmful pollution. Such rules valid for water protection zones are:

- Soil nutrient analysis, as well as corresponding fertilization plan is obligatory for all zones.
- Only in first zone: application of fresh manures or slurry, mineral N, use of fertigation, and plowing of permanent grassland are all prohibited.
- Only in third zone: first-quality sewage sludge use is acceptable.
- In all three zones: storage of manure or sewage sludge is prohibited.

Apart from what was already mentioned, there are also many other minor or locally valid pieces of legislation, but these major laws present the main structure of Slovenia's legislation on groundwater protection.

5. Slovenian water protective measures—the practice

Legislation is very important when it comes to protecting the environment, but without the knowledge of how to meet the prescribed criteria, stakeholders would have a hard time meeting it. This is where the more practical part of water protection comes in. Based on scientific research, trials, and experimenting, several instructive manuals with guidelines have been published for Slovenian farmers. Several institutions are a part of collective effort to discover locally efficient measures and practices in Slovenia; listed below are just a few:

- Biotechnical Faculty, University of Ljubljana;
- Agricultural Institute of Slovenia;
- Faculty of Agriculture and Life Sciences, University of Maribor;
- Chamber of Agriculture and Forestry of Slovenia;
- Biotechnological School Rakičan;
- Agricultural School Grm Novo Mesto;
- Biotechnical Center Naklo.

Collecting the most recent knowledge in the field, Ministry of Agriculture, Forestry and Food publishes (with advice from above-listed research institutions) a yearly issue of technological instructions for different agricultural sectors:

- integrated arable crops production;
- integrated fruit production;
- integrated grapevine production;
- integrated vegetables production.

Periodically, similar instructions are also issued for organic production. These manuals are not meant to address only the issue of groundwater protection, though. They are written to give farmers a framework for integrated, that is an environmentally friendly, sustainable production, which, as it happens, already helps in terms of groundwater protection. But—as stated in the beginning of this chapter, Slovenia has several areas, where just some generic measures are not sufficient. For this reason, more specialized guidelines are also issued to showcase the stricter solutions. Only recent publications are presented below, as older practices may no longer be relevant. A good representation of such a guidelines manual is "Fertilization with Nitrogen on Water Protection Areas (Example of Aquifer Apače Field)" [16] written in 2017. Important thing to stress out at this point is that only farms that are a part of either integrated or organic production scheme are certified and therefore under state control. In practice, this means that other farms, which are not a part of certified schemes and use conventional practices often, do not comply with legislation standards, because they are not stimulated or controlled enough to do so. Stimulation program exists, though, as a system of direct payments from European Union is available to farmers who enter into integrated or organic scheme, and only then, they enter into control mechanisms as well. Measures that need to be implemented in order to qualify for funding change every 6 years and have to do with animal welfare, cover-crops, good agricultural practices, water-sources protection, biodiversity, etc.

Authors of the manual note that for their example, area almost half of the farms in second and third vulnerable zone are not a part of official certified schemes, and therefore, are not subjected to control from state officials. Not that these farmers want to be polluters, sometimes they are just not willing to adapt to the new ways and just stick to what their forefathers taught them. Dealing with the lack of knowledge, practical tips for farmers that do not know how to deal with fertilizing plans and balanced fertilizing are presented in the manual. Some tips are listed below:

- When plowing a meadow (it is prohibited in zone 1), farmer should note that plant matter holds from 60 to 120 kg N/ha, and incorporate this quantity into his fertilizing plan.

- Soil samples for N analysis are best taken before first additional fertilization, and should be taken in regards to the size of the field. One sample is only enough for small fields up to 3 ha (though it should be taken from 20 to 30 random spots in the field), but for any size of field uniformity of soil and culture should be taken into account. With this, we provide for a uniform and representative sample.

- To minimize the chance of mineral N fertilizer leaching through the soil, we apply it in smaller rations: at seeding up to 60 kg/ha, and with additional rations only up to 80 kg/ha, but the total limit specified for different zones also should not be violated.

- Another good practice is to use a "fertilizing window," which is a small plot in the field where we apply less fertilizer than elsewhere. As plants use nitrogen, plants in the "window" will show signs of deficiency sooner, and additional fertilization can be planned accordingly. But if plants in the "window" grow fine, we know additional rations are not necessary.

- It is a good idea to cover at least 25% of nitrogen needs by either animal manures or by the use of legumes, green manures, or other cover crops, as nitrogen in these forms usually releases over longer periods.

- Manure or compost needs to be stored in dedicated sealed areas to prevent leaching.

- If possible, use of slow-release nitrogen fertilizers is encouraged.

- Application of fertilizer with implements that work it into the ground is preferential, to minimize both odor and volatilization, but also to get it close to the roots sooner.

Another work on fertilizing is the "Guidelines for realization of water protection claims in regards to nitrates from agricultural sources" [17] written in 2016, with updates from 2017. This is a very farmer-focused publication, written in a question-and-answer form, and is intended to explain legislation claims in a clear and easy-to-understand manner. Farmers are becoming more and more knowledgeable and eager for education, but long and complicated legislation pieces are not very easy nor enjoyable to read. Guidelines are not focused only on water protection areas, but rather include the good practices from legislation concerning the whole Slovenia. In many points, this is similar to the previous work, but at some it goes into more practical details, that farmers are interested in, like for example: "What is the animal-based equivalent of the yearly maximum of nitrogen

Farm animal	Yearly N content in manure (kg N)
Cattle	
Calves up to 6 months	10.5
Young cattle from 6 months to 1 year	21
Young cattle from 1 to 2 years	42
Cattle older than 2 years	70
Sheep	
Lambs (included in ewes content)	0.0
Ewes older than 1 year	10.5
Rams older than 1 year	10.5
Goats	
Baby goats (included in does content)	0.0
Does older than 1 year	10.5
Bucks older than 1 year	10.5
Horses	
Foals up to 1 year	30
Ponies	30
Horses older than 1 year	60
Donkeys and crossbreeds	30
Pigs	
Piglets (included in sows content)	0.0
Young pigs up to 30 kg	3.2
Pigs from 30 to 110 kg	11.2
Boars	27.2
Sows	25.6
Poultry	
Laying hens	0.420
Young hens	0.136
Broilers	0.170
Turkeys	1.700

Table 3.
Average yearly nitrogen yield from animal manure by species [17].

input (170 kg N/ha)?" A table with average yearly nitrogen yield for different species manure as well as a calculation example is included. Part of the table values is presented in **Table 3**. If farm animals produce more than 170 kg of nitrogen per hectare, excess manure should not be used on the farm, but sold or disposed elsewhere to prevent overloading of soil with nutrients.

Apart from this, other tips are included, for example, formulas to calculate:

- Nitrogen content in different ratio NPK fertilizers;
- Land inclination for purposes of surface water protection from erosion to streams;

- Size of storage area for manures depending on animal numbers on a farm;
- Nitrogen content in part of yield that is often times incorporated back into the soil (like straw) and acts as a nitrogen source when done so.

Manual also offers clarifications of legislation, like "Is it allowed to temporarily store manure on frozen ground?" (Yes), "Is anyone allowed to make a fertilizing plan?" (Yes, as long as he knows how to do it), "Is there a specific type of fertilizing implement supposed to be used?" (No, but those that incorporate the fertilizer into the ground are preferable), etc.

Moving from the topic of fertilizer pollution into the plant protection products territory, we have another recent (2017) manual on use of phytopharmaceutical substances in water protection areas [18]. At the beginning of the document, the author makes a very important statement about what it means to cause pollution: "Advances in measuring technology have made detection of plant protection products very sensible." Author gives us the following example: "When filling a spraying implement, a farmer drops the lid of a pesticide bottle, and it ends up in the nearby stream. Even if it only had a couple drops of pesticide on it, this can still be detected up to 40 km away in drought season, and even exceed the threshold values. If even such small mistakes or carelessness of a single user can be detected, it is not hard to imagine how important it is that every single user knows the principles of good agricultural practices, or his deeds can give bad results for whole area. The thinking in terms of 'this drop can't make any difference' or 'nobody will see me if I do it' is not helpful, as detection of pesticide in waters must, by law, be sanctioned by restrictive measures of some sort in the area of concern. This of course inflicts all the farmers in the area, not just the polluter."

In the manual, there is a good explanation of what is going on in the soil that is subject to leaching, but practical considerations are also included. Attached is the list of pesticides that are not acceptable or should only be used every couple years in water protection zones because of their chemophysical properties when in contact with soil (unacceptable active substances are bromacil, propaklor, triklorpyr, terbutryn, piloram, haloxyfop, terbacil, hexazinone, norflurazon, heptachlor, terbumeton). Just as with crop rotations, pesticide rotations should also be considered to minimize constant exposure to the same active substance, but also to allow for less frequent use of more dangerous pesticides in favor of using a similar but less hazardous active substance.

Another important topic is preparation of spraying mixture and dealing with its residuals after use. Care should be taken to prepare the right amount, a certified sprayer with good nozzles should be used (as discussed in the theory chapter), and a water-tight system for cleaning the sprayer should be used afterward. There is no urgent need to buy expensive devices, the author gives farmers an easy, cheap, and effective solution: If using a dedicated commercial cleaning system is not an option, smaller farmers can also create an improvised pool out of durable foil with a small quantity of soil in it—the soil works as a reactive substance in which, under sunlight, pesticide is decomposed. Another option is to empty the sprayer into the manure lagoon, where active substances also get decomposed quickly, especially with small quantities that remain after use. A big part of the document deals with surface drift into streams though, which is not our point if interest here.

6. Perspectives and improvements for future

Several projects on the topic of groundwater protection were recently concluded or are still in progress. Some of them are briefly described in the next lines.

"SI-MUR-AT" project [16, 17], dealing with ecological and sustainable agriculture in accordance to a contemporary water management is an interregional between Austria and Slovenia. The project is still underway, but one of the interesting interim propositions was to prepare a wholesome regulation on topic of proper cleaning of spraying equipment which is not set by Slovenian law yet. A good recommendation is also to give more emphasis to conservation management of soils, which in other countries shows promising results in terms of better soil properties. Also, just as crop rotation plans are mandatory for integrated or organic agriculture, plant protection products rotation plans would also be a good idea to have in order to gain a better view of the actual farming practices.

Another project, funded by Slovenian research fund and titled "Possibilities of farming in water protection areas" [19], offered some valuable insight into possible improvements of current state. Its aim was to discover measures to improve groundwater quality, and the final propositions were the following:

- Identification of the severely vulnerable aquifers and additional funding for farmers in these areas;
- GPS devices should be mounted onto fertilizing equipment in order to have a better control overfertilization in these areas;
- Redundant liquid manures should be composted in biogas plants or wastewater treatment plants to prevent overfertilization of vulnerable areas;
- Funding for building of small biogas plants, transportation of manure into them, and additional funding for farmers, willing to implement these measures.

"UraViVo" project [20], aimed to improve land management regime in order to improve groundwater quality, is another project that is still underway. Three most interesting goals are to develop and test a new formulation of fertilizer made from pig manure to minimize leaching possibilities, test a deficit irrigation practice to save water and minimize leaching possibilities and to use nitrate-polluted groundwater for fertigation in order to reduce nitrate concentration. Use of modeling to test more long-term results is also planned.

Workshops and lectures on topic of groundwater protection are also organized frequently by Chamber of Agriculture and Forestry of Slovenia's regional offices and also other institutions to extend the knowledge and make it available to farmers and interested public.

In the end though, even if scientists and researchers figure out good ways and effective measures, it is still on the legislators to legalize them, and on farmers or industry to implement them. Without the interest in implementation, even a very effective measure is not effective, as nobody implements it. Subsidies and funding seem to help in this regard significantly though.

7. Conclusions

Groundwater in Slovenia, as well as in many places across the globe, is the main source of drinking water. With the rapid increase in use of mineral nitrogen fertilizers and different agricultural as well as industrial chemicals in the past century, many of these substances found their way into groundwater. Slovenia as a European Union member formed its legislation based on Water Framework Directive and its recommendations.

Slovenia is not a big country, but it features very diverse terrain and climate. The main reservoirs of groundwater are located under alluvial plains, which are also centers of agricultural production. When speaking of groundwater pollution, leaching is generally the cause of it, especially on shallow sandy soils. Slovenia is no exception, as depth to groundwater is sometimes only 1–2 m. Apart from soil type, climate and land use also contribute to leaching, but land use is the only factor we can consciously change. To influence better use of land, legislation and guidelines are written.

Main pieces of groundwater-related legislation in Slovenia are the Waters Act and the Decree on groundwater status. They define rules for management of waters, vulnerable areas, threshold values, etc. Decree on the protection of waters against pollution caused by nitrates from agricultural sources, plant protection products act, and decree on the management of sewage sludge from the urban waste water treatment plants are also connected to groundwater protection in terms of definition of limitations, conditions of use, etc.

Many different guidelines were written in recent and past years to help realize the legislation expectations. They contain different instructions, good agricultural practices, lists of promising measures, as well as also very straightforward clarifications of legislation in Q&A form. In terms of the future perspectives, several projects driven by Slovenian research and education institutions are also presented in the work, and their efforts explained.

Author details

Miha Curk* and Matjaž Glavan
Department of Agronomy, Biotechnical Faculty, University of Ljubljana, Ljubljana, Slovenia

*Address all correspondence to: miha.curk@bf.uni-lj.si

References

[1] ARSO. Okolje se spreminja—Podnebna spremenljivost Slovenije in njen vpliv na vodno okolje. 2010. Available from: http://www.arso.gov.si/oagenciji/knjižnica/publikacije/Okolje_se spreminja.pdf [Accessed: October 11, 2018]

[2] Lam SK, Suter H, Mosier AR, Chen D. Using nitrification inhibitors to mitigate agricultural N_2O emission: A double-edged sword? Global Change Biology. 2017;**23**(2):485-489. DOI: 10.1111/gcb.13338

[3] Mosier AR, Syers JK, Freney JR. Nitrogen fertilizer: An essential component of increased food, feed and Fiber production. In: Agriculture and the Nitrogen Cycle: Assessing the Impacts of Fertilizer Use on Food Production and the Environment. Washington: Island Press; 2004. Available from: https://books.google.com/books?hl=en&lr=&id=W2UwatQ52TYC&pgis=1 [Accessed: October 11, 2018]

[4] Di HJ, Cameron KC. Nitrate leaching in temperate agroecosystems: Sources, factors and mitigating strategies. Nutrient Cycling in Agroecosystems. 2002;**64**(3):237-256. DOI: 10.1023/A:1021471531188

[5] Tribouillois H, Fort F, Cruz P, Charles R, Flores O, Garnier E, et al. A functional characterisation of a wide range of cover crop species: Growth and nitrogen acquisition rates, leaf traits and ecological strategies. PLoS One. 2015;**10**(3):e0122156. DOI: 10.1371/journal.pone.0122156

[6] Blesh J. Functional traits in cover crop mixtures: Biological nitrogen fixation and multifunctionality. Journal of Applied Ecology. 2018;**55**(1):38-48. DOI: 10.1111/1365-2664.13011

[7] Waters Act. Official Gazette of Republic of Slovenia. 2002. Available from: http://pisrs.si/Pis.web/pregledPredpisa?id=ZAKO1244 [Accessed: October 22, 2018]

[8] Rules on Determining Water Bodies of Groundwater. Official Gazette of Republic of Slovenia. 2005. Available from: http://www.pisrs.si/Pis.web/pregledPredpisa?id=PRAV6945 [Accessed: October 22, 2018]

[9] Rules on Determining and Classification for Water Bodies on Surface Water. Official Gazette of Republic of Slovenia. 2004. Available from: http://pisrs.si/Pis.web/pregledPredpisa?id=PRAV5957 [Accessed: October 22, 2018]

[10] Rules on Criteria for Marking a Water Protection Zone and a Bathing Water Zone. Official Gazette of Republic of Slovenia. 2005. Available from: http://pisrs.si/Pis.web/pregledPredpisa?id=PRAV6946 [Accessed: October 22, 2018]

[11] Decree on Groundwater Status. Official Gazette of Republic of Slovenia. 2009. Available from: http://www.pisrs.si/Pis.web/pregledPredpisa?id=URED5121 [Accessed: October 22, 2018]

[12] Environmental Protection Act. Official Gazette of Republic of Slovenia. 2005. Available from: http://pisrs.si/Pis.web/pregledPredpisa?id=ZAKO1545 [Accessed: October 24, 2018]

[13] Decree on the Protection of Waters Against Pollution Caused by Nitrates from Agricultural Sources. Official Gazette of Republic of Slovenia. 2009. Available from: http://www.pisrs.si/Pis.web/pregledPredpisa?id=URED5124 [Accessed: October 24, 2018]

[14] Plant Protection Products Act. Official Gazette of Republic of Slovenia. 2012. Available from: http://pisrs.si/Pis.

web/pregledPredpisa?id=ZAKO6355 [Accessed: October 24, 2018]

[15] Decree on The Management of Sewage Sludge From The Urban Waste Water Treatment Plants. Official Gazette of Republic of Slovenia. 2008. Available from: http://www.pisrs.si/Pis.web/pregledPredpisa?id=URED4880 [Accessed: October 24, 2018]

[16] Bavec F. Gnojenje z dušikom na vodovarstvenih območjih (Primer vodonosnika Apaškega polja) [report]. University of Maribor, Faculty of Agriculture and Life Sciences Hoče; 2017. Available from: http://www.kgzs-ms.si/wp-content/uploads/2017/09/Gnojenje-z-N-na-VVO.pdf [Accessed: October 20, 2018]

[17] Sušin J, Verbič J, Matoz H. Smernice za izvajanje zahtev varstva voda pred onesnaže vanjem z nitrati iz kmetijskih virov [report]. Ministry for Environment and Spatial Planning; 2017. Available from: http://www.mop.gov.si/fileadmin/mop.gov.si/pageuploads/podrocja/voda/smernice_2017.pdf [Accessed: October 26, 2018]

[18] Lešnik M. The Use of Plant Protection Products and the Protection of Water at Water Protection Zones. University Press University of Maribor; 2017. DOI: 10.18690/978-961-286-030-1

[19] Urbanc J. Možnosti kmetovanja na vodovarstvenih območjih [report]. Ljubljana: Geological Institute of Slovenia; 2017. Available from: https://www.dlib.si/details/URN:NBN:SI:DOC-HWYIJSVV/ [Accessed: October 29, 2018]

[20] Pintar M. UraViVo—Project Presentation [Leaflet]. 2017. Available from: http://www.bf.uni-lj.si/index.php?eID=dumpFile&t=f&f=30765&token=3e80d3c2e2db2e5902156bf336f05a487fd243ac [Accessed: October 29, 2018]

Chapter 7

Groundwater Management Competitive Solutions: The Relevance of the Gisser-Sanchez Model

Oscar Alfranca

Abstract

The main subject of this chapter is related to the relevance of the Gisser-Sanchez effect in groundwater. It is important to point out that groundwater resources provide a primary source of irrigation water throughout much of the world. Two main questions need to be indicated when taking water extractions into account. The first has to do with water scarcity in local watersheds or whole basins created by excessive surface and groundwater withdrawals. The other is related to water degradation and the pollution loads leading to many tracts of rivers and whole aquifers being spoiled and losing their capacity to sustain ecosystem functioning and human activities. These conclusions were called into question by the Gisser and Sanchez analysis. These authors argue that the difference in producer surplus between the open access and optimally managed cases was numerically insignificant for large aquifers subject to inelastic water demand. Perhaps the most interesting point in the work by Gisser and Sanchez is multidisciplinarity.

Keywords: Gisser-Sanchez effect, groundwater, property rights

1. Introduction

Groundwater management is an issue which remains a practical matter in many human regions throughout the world [1]. Besides, it is very necessary to clarify that groundwater represents the largest stock of accessible freshwater and accounts for about one-third of freshwater withdrawals globally [2–4]. However, increased rainfall scarcities have resulted in an augmented use of groundwater, in order to satisfy the increasing domestic, agricultural, and environmental-ecosystem preservation for different water.

Nevertheless, it is necessary to take into account that historically, surface water has been the main source of water for human consumption, as it was easy and cost effective to access. So, it can be expected that during the second half of the twentieth century, groundwater withdrawals will increase. It is also very relevant to reflect that groundwater supply could represent around one third of the world population [5].

This wide use of groundwater in many parts of the world has resulted in water level decline and groundwater depletion and is mainly related to phenomena such as biodiversity loss, pollution, and seawater intrusion in coastal aquifers.

An example could be found in the paper by El Moujabber et al. [6], in which the state of groundwater desalination by seawater intrusion in the Lebanese cost is introduced (specifically in the region of Choueifat-Rmeyle, located in the south of Mount-Lebanon). The main consequence that is obtained is related to the fact that groundwater management can behave like relevant backstop technologies and also that substitutes have become a practical concern in many arid and semiarid regions throughout the world [7].

A fundamental idea that needs to be pointed out is that groundwater is essential for sustaining agriculture production patterns, as well as consumption models and the biodiversity or the resilience of ecosystems. The combination of this fact with the intense scarcity in many parts of the world makes necessary the development of rules for the corrected and efficient allocation of resources among competing uses over time and space.

This presents an economic question which has been close to groundwater economics since the middle years of the decade of 1950s. It is necessary to point out that the question of how to manage this resource, mainly because groundwater constitutes about 89% of the freshwater on earth (discounting that in the polar ice caps). From this, an important economic concept could be deduced related to water scarcity and which is related to the fact that the world water scarcity is one of the most important hydraulic resources that need to be taken into account.

It is also necessary to point out that groundwater systems are rather dynamic with groundwater in motion from zones of recharge to areas of discharge and that a great number of years could, hundreds of years, interfere in the passage of water through this subterranean part of the hydrological cycle. Since flow rates regularly do not ordinarily go beyond a small number of meters per day and can be as low as 1 meter per year (these groundwater velocities compare to rates of up to 1 meter per second for river flows) [8].

Groundwater resources provide a primary (or supplemental) source of irrigation water throughout much of the world, yet overpumping and subsequent aquifer depletion may pose "the single largest threat to irrigated agriculture" [9, 10].

Two main questions need to be indicated in when taking water extractions into account. The first is double: one is water scarcity in local watersheds (or whole basins created by extreme surface and groundwater withdrawals). The other is water degradation from pollution loads leading to many tracts of rivers and whole aquifers being damaged and losing their capacity to sustain ecosystem functioning and human accomplishments.

Following the wide-scale development of groundwater pumping for agriculture in 1950s, some results have been obtained that the open access nature of groundwater implied that farmers were overextracting water, and therefore, it could be exhausted much before than it might be economically optimal.

These conclusions were called into question by Gisser and Sanchez. These authors, mainly in their very influential paper argue that the difference in producer surplus between the open access and optimally managed cases was numerically insignificant for large aquifers subject to inelastic water demand. Perhaps the most interesting point in the work by Gisser and Sánchez is multidisciplinarity. An essential assumption that we need to take into account is that the GSE model is a dynamic model. Besides, we need to take into account that variables in the model are economic, hydrological, and agronomic variables of groundwater use. In this chapter, the demand and supply functions for irrigated water are defined, and these functions are associated with the hydrological characteristics of the aquifer. Then, the path of water allocation through time is calculated under the policy regime and the free-market regime [11].

This effect has remained controversial, and numerous studies have analyzed whether the Gisser and Sanchez Effect (GSE) persists under a variety of specific conditions, such as convex pumping costs [12], shifting (nonconstant) water demand [13], adaptation by crop shifting [14], confined aquifers [15], heterogeneous users [16]; strategic decision-making [17, 18], conjunctive management [19], risk aversion [20], and backstop water sources [21]. These studies generally find support for the GSE, even under all these different conditions.

Nevertheless, it is necessary to point out that authors, such as Stratton et al. [22], apply the GSE model, although relaxing a very important significant assumption of a fixed irrigation technology. Results indicate that the GSE fails when irrigation technologies with different water use efficiency become available. These results are robust and hold even when maintaining some of the very fundamental statements in the original model (such as constant marginal pumping costs per linear foot of lift). Besides, the gains from optimal groundwater management become even more significant when irrigation technology is not only variable but also endogenous variables. That is, variables whose values the model is designed to explain. In the model, there are also exogenous variables. That is, variables whose values are taken as given from outside the model [23]. The expression "Endogenous Technical Change" implies that higher water costs could induce the development of technologies that might improve water use efficiency [19, 24]. The expression "Endogenous Technical Change" implies that higher water costs could encourage the development of technologies that might improve water use efficiency [24, 25].

The main objective of this chapter is to re-evaluate the validity of the GSE hypothesis in groundwater management. In this chapter, the conceptual framework within which the elements interacting in the management of groundwater resources is examined. The most important conclusion obtained is that the role of the market is limited with respect to the price of water in an aquifer. This is an important result, because it points to the mechanism that could pull competitive water prices and quality-graded quantity of groundwater, in line with their equilibrium levels. In Section 2, some models of groundwater use and management are introduced, and the most important economic models for groundwater use can be found (joint with the potential of groundwater management control variables in such models). In Chapter 3, some relationships between the Gisser and Sanchez effect and the difficulties to establish clear groundwater property rights are discussed. In Section 4, the robustness of GSE under a private if property rights regime is discussed, both in quantity and in quality terms. In Section 5, a discussion section is introduced. Finally, some conclusions are provided.

2. Some models of groundwater use and the potential for groundwater management

It is necessary to take into account that implicit in the different concerns about groundwater, an essential principle can be found. This is related to the fact that if no intervention exists, then groundwater pumping will be mismanaged. Another important point that needs to be pointed out is that if groundwater pumping is inefficient, then, the lack of central (and optimal), control, underlines that the estimates of the welfare loss (under the common property regime) should depend on the specific model of firm behavior which might be enlisted in the analysis. This should allow to conclude in favor of an existing potential and pressing need for the development and implementation of management policies for groundwater resources [32].

It is also interesting to point out that when groundwater withdrawals exceed recharge, water will be mined over time until either supplies are exhausted or the marginal cost of pumping additional water should become extremely expensive [33]. An essential issue related to this assumption is that a marginal user cost is associated with mining groundwater, and this is related to the opportunity cost which is connected with the unavailability in the future of any unit of water used in the present.

A well-organized distribution should consider this user cost, which effectively signals the scarcity of the resource and is called the resource's scarcity rents. Therefore, efficient pricing of a resource that exhibits natural supply constraints incorporates both marginal cost of extraction and scarcity rents. Scarcity rents must be imposed on current users.

Given the complexity of establishing clear groundwater property rights, scarcity rents are frequently difficult to be recognized and are not easy to be estimated. Some authors in which a discussion about this point could be found are, for instance [31, 34–37].

Ignoring scarcity rents implies that the price of groundwater is usually too low and extraction is above the socially optimal level. If an optimal dynamic management of common-pool groundwater resources is not considered, or in the presence of a competitive extraction regime ignoring scarcity rents, results in inefficient pricing and misallocation of resources. This essential argument has to do with the way markets behave, and it could perfectly be competitive. Under these circumstances, the problem is not so much with the market mechanism but with the way property rights behave.

3. Groundwater property rights

Given the difficulty of establishing clear groundwater property rights, scarcity rents are frequently difficult to be estimated. Ignoring scarcity rents should imply that groundwater prices could be too low and extraction might be above the socially optimal level. From this, the main conclusion is that, in the absence of optimal dynamic management of common pool groundwater resources, or, alternatively, in the presence of a competitive extraction regime, ignoring scarcity rents, could result in inefficient pricing and misallocation of resources.

Just in the case, there is no optimal dynamic management of common pool groundwater resources, or, alternatively, in the presence of a competitive extraction regime, ignoring scarcity rents results in inefficient pricing competitive extraction regime, inefficient pricing, and misallocation of the resource.

From this, an interesting question might be pointed out: How could be explained that a competitive dynamic solution of groundwater exploitation is almost identical (in terms of derived social welfare) to the efficient management solution, in the way it is claimed by the GSE effect?

3.1 The Gisser-Sánchez effect

The GSE explains a contradictory empirical result, present and persisting in the dynamic solutions of groundwater exploitation under different extraction regimes (since 1980) [1]. In spite of the fact that depletion of aquifers is a major threat to many freshwater ecosystems all over the world, the social benefits from managing groundwater are numerically insignificant. It needs to be pointed out that GSE encompasses to a general rule, and then the role and scope of water management are severely limited. It is also essential to point out that, even if implementing optimal extraction is not going to be costless. In this section, a review of [38] is introduced about the theoretical and empirical attempts to address the GSE and discuss the potential for groundwater management.

3.2 The Gisser-Sanchez model and groundwater management

Problems of groundwater allocation have been studied basically in the context of the theory of mine [26–29]. The basic model by Gisser and Sanchez is a simplified representation of the economic, hydrologic, and agronomic facts that must be considered relative to the irrigator's choice of water pumping [1]. The validity of the GSE model rests on the key assumption that the aquifer has to be quite large and on the secondary assumption of a small slope in the water-demand function.

A separate literature should also have to be taken into account, which deals with groundwater quality. Some papers in this line can be found such as [30–32].

Groundwater allocation problems have been studied mainly in the context of mine and economists like [33–35]. Some principles of inventory management to derive decision rules for the optimal temporal allocation in a dynamic programming format can also be found in such papers. The effects of different policy instruments that could correct misallocation of commonly owned groundwater can be found in papers such as [31, 35–39], which studied the effects of different policy instruments that might correct the misallocation of commonly owned groundwater. One of the main results of this chapter is that net benefits from groundwater management could amount to over $100 per acre, but noted that these benefits could decline with increases in interest rate. One of the solutions to this problem was obtained by authors such as Allen and Gisser [40], who derived a formula for a tax that should be imposed on groundwater which was pumped in order to yield the optimal control solution. Finally, in papers such as [41], it can be recognized the issue of congestion externality in aquifers with open access characteristics and suggested a charging tax to accommodate this externality.

When this point is achieved, farmers will either import supplemental water or be restricted to use a smaller amount of water by being assigned water rights. Nevertheless, some changes in the hypothesis related to regulation of water pumping in the aquifer could be made. This case allows to model consistently an optimal control problem and also allows one kind of clarification that should be related with the case of no control. This is the departure point for the works [9, 10] by Gisser and Sanchez.

The basic model analyzed by Gisser and Sanchez is a simplified representation of the economic, hydrologic, and agronomic facts that should be considered for the irrigator's choice of water pumping. An irrigator benefit function could be represented using this function suggested by [44]:

$$\pi(t) = V((wt)) \quad C(H(t))w(t) \tag{1}$$

where $\pi(t)$ denotes profits at time t. Net farm revenue from water use $\pi(t)$ neglecting pumping costs is denoted by

$$V(w) = \int_0^w p(x)dx \tag{2}$$

where $p(x)$ is the inverse demand function for water. $C(H)$ is the average and marginal pumping costs per acre-foot of water and $H(t)$ is the height of water table above some arbitrary reference point at time t [1, 40]. The change in the height of water is given by differential Eq. (2), which represents the hydrologic state of the aquifer (or equivalently, the environmental constraint of the problem)

$$\acute{H} = \frac{1}{AS}(R + (a - 1)w), H(0) = H0 \tag{3}$$

In this equation, R exemplifies a constant recharge determined in acre feet per year; a is the constant return flow coefficient (which could be considered to be just a simple number); H_0 is the initial level of the water table measured in feet above sea level; A is the surface area of the aquifer (uniform at all depths), measured in acres per year; and S is the specific yield of the aquifer. These equations are based on the UNESCO-Encyclopedia Life Support Systems and also on the papers by [1, 43, 45] on the Gisser-Sanchez effect.

More precisely, the aquifer in Gisser and Sanchez's work is modeled as a bathtub, unconfined aquifer, with infinite hydraulic conductivity. It is necessary to point out that infinite hydraulic conductivity implies that the aquifer will never dry up, irrespective of groundwater extraction rates, which is equivalent to the assumption of a bottomless aquifer. The adoption of this hypothesis can be acknowledged by the hypothesis that it is implied by an standard hypothesis which is related to the literature and which implies that time goes to infinity [1]. Nevertheless, if this is not this way, a steady-state solution might not be reached. Besides, Provencher [43] showed that the optimal pumping rate can be substantially lower when the hydraulic conductivity is small enough to result in a significant cone of depression around the well. The assumption of constant return flow in the presence of fixed irrigation technology suggests a constant rate of water application.

The hypothesis of deterministic and constant recharge in conjunction with the hypothesis of constant return flow suggests constant types of land use [44], independence of surface water and groundwater systems, and constant average rainfall. Besides, sunk costs, replacement costs, and capital costs in general are overlooked, and it is implicitly assumed that energy costs are constant. It is also indirectly accepted that the well pump capacity constraint is nonbinding. Finally, refinement in Gisser and Sanchez's model could be also achieved by assuming that only land superimposing the aquifer can be irrigated. That is, the demand curve does not shift to the right over time. This implies that, the unambiguous recognition of the fact that the main hypothesis behind the GSE indicates that the result should be carefully when working on real aquifer systems.

Given the above hydroeconomic model, Gisser and Sanchez used a linear water demand function (estimated by [31, 32]) using parametric linear programming, hydrologic parameters that were considered realistic in the 1960s, and a discount rate of 10%, and simulated the intertemporal water pumpage for Pecos Basin in New Mexico, once under the assumption of no control and once under the assumption of optimal control. The most interesting result is that the trajectories under the two regimes are almost identical. This result leads to the main conclusion that there is no substantive quantitative difference between socially optimal rules for pumping water and competitive rates. Therefore, the welfare loss from intertemporal misallocation of pumping effort is negligible. This conclusion amounts to the GSE.

An important effect to consider is that, solving analytically the model, Gisser and Sanchez main result is that, if Eq. (3) is true, then the difference between the two strategies is so small that it can be ignored for practical consideration, where Eq. (3) is

$$\left[\frac{k\,C_t(a-1)^2}{AS}\right] \simeq 0 \tag{4}$$

In Eq. (4), k can be considered to be the reduction in demand for water per $1 intensification in price (that is, the slope of the uncompensated demand curve for groundwater), C_t is the intensification in pumping cost per acre-foot per 1-foot decline in the water table, and AS are given in Eq. (2). If Eq. (3) holds, then the rate of discount will be practically identical with the exponent of the competition result. Therefore, as long as the slope of the groundwater demand is small relative to the aquifer's area times its storativity [1], GSE will persist. From this, the main conclusion is that, if differences between optimal and competitive rates of water pumping are small, then policy considerations can be limited to those which ensure that the market operates in a competitive fashion, and concerns relative to rectifying common property effects could be removed.

3.3 Robustness of the GSE effect

The GSE effect presents important policy implications. Some empirical papers discussing the robustness of this effect are, Noel et al. [35] found that control increases the value of groundwater in the Yolo basin in California, by 10%. This result is fairly different from [37], who found that control raised the net benefit of groundwater in the Ogallala basin by only 0.3% empirical estimates of benefits from groundwater management in Kern county (California, USA) do not exceed 10%. Nevertheless, in works such as [39], it can be found that groundwater management in the Texas High Plains would be unwarranted, and he proceeded with a sensitivity analysis of present value profits using different slopes and intercept values for the groundwater-demand curve. It is interesting to point out that this analysis indicated that benefits from groundwater management do not increase monotonically as the absolute value of the slope increases.

A basic hypothesis of the Gisser and Sánchez model is that the demand curve for water is linear. This is a fairly conventional hypothesis in most economic demand models. In order to study the relative importance of this hypothesis for the GSE, optimal control and no-control strategies are compared, using a nonlinear demand curve [40]. This comparison confirmed that, for the case of the nonlinear demand function, what had been demonstrated by the GSE for the case of a linear demand function.

However, in works such as [20], it can be found that the differences between the two regimes may not be trivial if the relationship the average extraction cost and the water table level and/or if there exist significant differences in land productivity, applying dynamic programming to a model of a confined aquifer underlying the Crow Creek Valley in South-Western Montana.

It is essential to take into account that when land is assumed to be homogeneous, the gross returns function with respect to water use tends to be nearly linear. Nevertheless, with greater heterogeneity in productivity, the returns function is more concave, and differences in the optimal use policy under a common property setting are more pronounced [1]. Hence, the need for more theoretical work is to determine an asymmetric groundwater pumping differential game, where differences in land productivity are taken into account.

3.4 Variable relations and endogenous rates of change

Implicit in GSE model is the hypothesis of nonvariable economic relations (that is, time-independent demand) and/or exogenous and constant rates of change (that is, constant and fixed exogenous crop mix, constant crop requirements, fixed irrigation technology), and some significant exceptions can be found such as [43, 44], with constant exogenous kinds of land use and nonvariable hydrologic conditions.

Nevertheless, in studies with a long run perspective, predictable results could turn out to be weaker as the steady state is approached. Estimated benefit and cost functions used in the simulations of GSE may bear little relation to the actual benefit and cost functions when economic, hydrologic, and agronomic conditions are much different. More complex representations of increasing resource scarcity incorporate opportunities for adaptation to the rising resource prices which are a main indicator for scarcity. In the long run, adoption of new techniques, substitution of alternative inputs, and production of a different mix of products offer rational responses to increasing scarcity [1], [38].

4. The robustness of GSE under a private property rights regime

The solution which is commonly proposed for the inefficiencies arising in common property resource extraction is central-optimal control by a regulator, who uses taxes or quotas to obtain the efficient allocation of resources over time.

In the background of groundwater depletion, a solution has been commonly suggested which is based on a tradable permit scheme [37, 38]. In the framework of groundwater reduction, a number of authors have recommended a similar institutional arrangement in which firms are arranged and endowment of tradable permits to the in situ groundwater stock, which they control over time. Each firm's bundle of permits represents its private stock of groundwater.

This private stock is worsening due to groundwater pumping and intensifications to reflect the firm share of periodic recharge. It also changes in response to the activity of the firm in the market for groundwater stock permits, increasing when permits are purchased and decreasing when permits are sold. The market price for permits serves to allocate groundwater over time.

It is necessary to point out that this particular regime is inefficient, mainly because both the pumping cost externality and the risk externality persist after the allocation of permits. Moreover, this regime is time inconsistent. However, different efforts to quantify the value of groundwater resource under both optimal control and the private property rights regime indicate that groundwater privatization recovers most of the potential gain from management. In particular, a programming model for Madera County, in California (USA), can be found in [37]. This regime recovered 95% of the potential gain from management.

4.1 The GSE in models of conjunctive use of surface and groundwater

A tributary aquifer is characterized by a groundwater stock that is hydrologically connected to a body of surface water. In this aquifer, surface water may recharge the underground aquifer, or groundwater may supplement surface flows depending upon hydrological conditions.

In papers such as [38], results can be found in which an analytical economic model is developed and is focused primarily on the hydrologic link between surface and groundwater, by modeling the instantaneous rate of aquifer recharge caused by groundwater pumping, through river effects. In this chapter, some externalities river effects can be found, which reinforced groundwater overpumping present due to the usual common property effects. Results of this chapter indicate that optimal policy requires compensation to be paid for both river effects and aquifer depletion net of river effects. This work points to an externality created by groundwater overpumping provoked mainly by the common property effects.

From this, the main conclusion which needs to be pointed out is that optimal policy requires a recompense to be paid for both river effects and aquifer depletion net of river effects [39]. It is necessary to highlight that these effects indicate the existence of some externalities which could be related to groundwater pumping, which might be adjusted with the precise management. The main consequence probably could be that GSE might be very likely removed by the improvement in management benefits.

Unfortunately, no empirical results exist of these results focusing primarily on the hydrologic link between ground and surface water, and at the same time acknowledging the stochastic nature of surface water supplies. Instead, the main literature that incorporates stochastic surface supplies into a groundwater model in which surface water and groundwater are modeled as substitute goods, aquifers are not connected with surface water, and they only benefit from substantial natural recharge.

5. Discussions

Regarding the GSE model, it needs a number of important assumptions. One of the most significant has to do with the disregard for aquatic ecosystems linked and dependent on aquifer systems.

In the GSE model, a very special point needs to be pointed out, which is that the aquifer is presented as a "bath-tub", unconfined aquifer, with infinite hydraulic conductivity [40]. A bath-tub approach to modeling an aquifer assumes that it responds uniformly and instantly to groundwater extraction [41, 46, 47]. From this, the spatial distribution of the users of the resource is not so relevant, and the evolution of the spatial profile of drawdown does not affect current and future extraction choices. Gisser-Sánchez assumes a deterministic and constant recharge, constant return flow and average rainfall, independence of surface water and groundwater systems, and a bottom-less aquifer. Since their competitive steady state presents a positive water stock, their estimation of welfare gains from optimal management excludes stock externality [42].

Another important assumption which is discussed in this chapter is the appropriateness of the stock effect assumption. This hypothesis reflects the dependence of extraction costs and the eventual benefits on the stock of the resource. From this, it could be established the way these assumptions might affect the time variation of the shadow price of groundwater externality. In this chapter, the main result is that this could lead to a declining value of in situ resource over time. Therefore, the addition on nonmarginal extraction costs could be close to inappreciable, which could imply the validation of the Gisser-Sanchez effect, which also presents the remarkable hypothesis that groundwater markets have the benefit of allowing more flexible movement of water to serve changing conditions and demands.

Finally, a main conclusion could be derived from the paper introduced which is that a very relevant model such as [4, 5] is a very appropriate work to analyze groundwater management, mainly because it states the conditions under which welfare improvements from policy interventions could be significant in aquifer administration. This result could be compared with nonregulation or free market solutions in groundwater management.

6. Conclusion

The main conclusion of this chapter has to do with the GSE effect and points mainly to the different effects related to welfare improvements and aquifer management. In this work, an optimal policy requires a compensation to be paid for both river effects and aquifer depletion, which points to an additional externality created by groundwater pumping. This externality could be corrected with an appropriate management of the groundwater, which could eventually eliminate the GSE effect and even increase management benefits.

No empirical results have been obtained in order to test these results, which have to do mainly with the eventual links between ground and surface water. These results could be pertinent in order to improve groundwater management, because from this, the stochastic nature of surface water flows could be acknowledged.

Nevertheless, probably the most significant result in this chapter is that different effects related to welfare improvements and aquifer management and the relevance of the GSE effect exists. Besides, it is necessary to indicate that an optimal aquifer management policy requires a compensation to be paid for both the existing river effects and aquifer depletion. These conclusions stem from the fact that externalities exist, which are linked to groundwater pumping. This externality could also be corrected with a suitable management of the groundwater. This result is quite relevant because it could potentially remove the GSE effect, and therefore, even increase management benefits.

It is similarly essential to take into account the appropriateness of some of the assumptions in the model, since some of them (like the linear relationship between pumping costs for nonconsumptive benefits), and which are an essential tool in groundwater management.

Environmental uses of groundwater water and the way markets work present a significant impact on users and the environment. An interesting conclusion is provided by [48], in his paper for the journal *Resources* (from Resources for the Future), which is that there is rapid depletion of aquifers in the United States, and this presents significant impacts on users and the environment, requiring stakeholders across the country to look for creative and effective policy solutions. So, there is an interesting conclusion that groundwater markets can be applied broadly in groundwater management in order to protect one the most relevant freshwater environmental resource.

Author details

Oscar Alfranca
Departament d'Enginyeria Agroalimentaria i Biotecnologia, Universitat Politècnica de Catalunya, Barcelona, Spain

*Address all correspondence to: oscar.alfranca@upc.edu

References

[1] Koundouri P. Current issues in the economics of groundwater resource world management. Journal of Economic Surveys. 2004;**18**(5):703-740

[2] Gorelick SM, Zheng C. Global change and the groundwater management challenge. Water Resources Research. 2015;**51**:3031-3051

[3] Farmigletti JS. The Global groundwater crisis. Nature Climate Change. 2014;**4**:945-948

[4] Siebert S, Burke J, Faures JM, Frenken K, Hoogeveen J, Doll P, et al. Groundwater use for irrigation. A global inventory. Hydrology and Earth System Sciences. 2010;**14**:1863-1880

[5] Oki T, Kanae S. Global hydrological cycles and world water resources. Science;**313**:1068-1072

[6] El Moujabber E, Bou Samra B, Darwish T, Atallah T. Comparison of different indicators for groundwater contamination by seawater intrusion on the Lebanese coast. Water Resources Management. 2006;**20**(2):161-180

[7] Momejian N, Abou Najm M, Alameddine I, El_Fadel M. Can groundwater vulnerability models assess seawater intrusion? Environmental Impact Assessment Review. 2019;**75**:13-26

[8] Ge S, Gorelick SM. Groundwater and surface water. In: Encyclopedia of Atmospheric Sciences. 2nd ed. Oxford, Elsevier, Academic Press

[9] Gisser M, Sánchez DA. Some additional economic aspects of groundwater resources replacement flows in semi-arid agricultural areas. International Journal of Control. 1980a;**31**(2):331-341

[10] Gisser M, Sánchez DA. Competition vs optimal control in groundwater pumping. Water Resources Research. 1980b;**31**:638-642

[11] Esteban E, Albiac J. Groundwater and ecosystem damages: Questioning the Gisser-Sanchez effect. Ecological Economics. 2011;**70**:2062-2069

[12] Negri D. The common property aquifer as a differential game. Water Resources Research. 1989;**25**(1):9-15

[13] Rubio S, Casino B. Competitive versus efficient extraction of a common property resource: The groundwater case. Journal of Economic Dynamics and Control. 2001;**25**(8):1117-1137

[14] Provencher B, Burt O. The externalities associated with the common property exploitation of groundwater. Journal of Environmental Economics and Management. 1993;**24**:139-158

[15] Knapp KC, Olson L. The economics of conjunctive groundwater management with stochastic surface supplies. Journal of Environmental Economics and Management. 1995;**28**:340-356

[16] Koundouri P, Christou C. Dynamic adaptation to resource scarcity and backstop availability: Theory and application to groundwater. Australian Journal of Agricultural and Resource Economics. 2006;**50**:227-245

[17] Kim H, Kaown D, Mayer B, Lee J. Identifying the sources of nitrate contamination of groundwater in an agricultural area (Haean basin, Korea) using isotope and microbial community analyses. Science of the Total Environment. 2015;**533**(15):566-575

[18] Brill TS, Burness HS. Planning versus rates of groundwater pumping. Water Resources Research. 1994;**30**:1873-1880

[19] Kim CS, Moore MR, Hanchar JJ. A dynamic model of adaptation to resource depletion: Theory and an application to groundwater mining. Journal of Environmental Economics and Management. 1989;**17**:66-82

[20] Worthington V, Burt O, Brustkern RL. Optimal management of a confined groundwater system. Journal of Environmental Economics and Management. 1985;**12**(3):229-245

[21] Saak AE, Peterson JM. Groundwater use under incomplete information. Journal of Environmental Economics and Management. 2007;**54**(2):214-228

[22] Stratton SE, Simon LK, Marchiori C. Promoting groundwater reform in the Guadiana Basin. American Journal of Agricultural Economics. 2008;**90**:1343-1349

[23] Hoy M, Livernois J, McKenna C, Rees R, Stengos T. Mathematics for Economics. Cambridge, Massachusetts: The MIT Press; 2001

[24] Kim CS, Guilfoos T. The effect of cost-share programs on groundwater exploitation and nonpoint-source pollution under endogenous technical change. Agricultural and Resource Economics Review. 2016;**45**(2):394-417

[25] Kim CS, Fuglie K, Wallander S, Wechser S. Endogenous Technical Change and Groundwater management: Revisiting the Gisser-Sanchez Paradox. AAEA and EAEA Joint Meeting; July 26-29. San Francisco, CA; 2015

[26] Provencher B, Burt O. A private property rights regime for the commons: The case for groundwater. American Journal of Agricultural Economics. 1994;**76**:875-888

[27] Groom B, Koundouri P, Swanson T. The watershed economics management approach: An application to Cyprus. In: Koundouri P, Pashardes P, Swanson T, Xepapadeas A, editors. Economics of Water Management in Developing countries: Problems, Principles and Policies. Cheltenham, UK: Edward Elgar; 2003a

[28] Groom B, Koundouri P, Swanson T. Integrated resource management in Cyprus: Introducing and applying a new methodology. In: Brower R, Pearce D, editors. The Economics of Water. Dordrecht, The Netherlands: Kluwer; 2003b

[29] Knapp K, Barenklau K. Groundwater quantity and quality management: Agricultural production and aquifer salinization over long time scales. Journal of Agricultural and Resource Economics. 2006;**31**:616-641

[30] Milliman JW. Commonality, the price system and the use of water supplies. Southern Economic Journal. 1956;**22**:426-437

[31] Gisser M, Mercado A. Integration of the agricultural demand functions for water and the hydrologic model of the Pecos Basin. Water Resources Research. 1972;**8**:1373-1384

[32] Gisser M, Mercado A. Economic aspects of groundwater resources and replacement flows in semiarid agricultural areas. American Journal of Agricultural Economics. 1973;**55**:461-466

[33] Cummings R, McFarland JW. Groundwater management and salinity control. American Journal of Agricultural Economics. 2003;**56**:830-836

[34] Silveira L. Special issues in Groundwater. In: Silveira L, Usunoff E, editors. Groundwater, vol. III. Eolss Publishers Company Limited; 2009

[35] Noel JE, Gardner BD, Moore CV. Optimal regional conjunctive water management. American Journal of Agricultural Economics. 1980;**62**:489-498

[36] Gisser M, Mercado A. Economic aspects of groundwater resources and replacement flows in semiarid agricultural areas. American Journal of Agricultural Economics. 1973;**55**:638-642

[37] Lee KC, Short C, Heady EO. Optimal groundwater mining in the Ogallala Aquifer: Estimation of economic losses and excessive depletion due to commonality. In: Center for Agricultural and Rural Development. Iowa, USA: Iowa State University; 1981

[38] Das S. The economics of groundwater resource management. International Journal of Ecosystems. 2015;**5**(3A):65-68

[39] Nieswiadomy M. The demand for irrigation water in the High Plains of Texas. American Journal of Agricultural Economics. 1985;**67**:619-626

[40] Allen RC, Gisser M. Competition versus optimal control in groundwater pumping when demand is nonlinear. Water Resources Research. 1984;**20**:752-756

[41] Burness H, Brill TC. The role for policy in common pool groundwater use. Resource and Energy Economics. 2001;**23**:19-40

[42] Shah FA, Zilberman D, Chakravorty U. Technology adoption in the presence of an exhaustible resource: The case of groundwater extraction. American Journal of Agricultural Economics. 1995;**77**(2):291-299

[43] Provencher B. Issues in the conjunctive use of conjunctive use of surface water and groundwater. In: Bromley DW, editor. The Handbook of Environmental Economics. Oxford, UK/Cambridge, MA: Blackwell; 1995

[44] Reddy V. Costs of resource depletion externalities: A study of groundwater overexplotation in Andhra Pradesh, India. Environment and Development Economics. 2005;**10**:533-556

[45] Tietenberg T, Lewis L. Environmental Economics & Policy. 10th ed. London: Pearson; 2018

[46] Katic PG. Three essays on the economics of groundwater extraction [Ph.D. thesis]. Australia: The Australian National University; 2011

[47] Katic PG, Grafton RQ. Economic and spatial modelling of groundwater extraction. Journal of Hydrology. 2012;**20**:831-834

[48] Kuwayama Y. Groundwater markets: Managing a critical, hidden resource, resources. Resources for the Future. 2014;**43**:1-11